SABENA

The Airline with The Curvy 'S'
1923-2001

Theodore Gianna

CLASSIC AIRLINES PUBLICATIONS

Front cover photo by Norbert Kröpfl

ISBN: 978-0-6489379-5-1 (Paperback)
First printing edition 2022.

TABLE OF CONTENTS

INTRODUCTION

Sabena began operations in 1923. The Belgian flag carrier extended its route network further and further each year from then on. In the early years, the carrier's prize route was Brussels to Leopoldville, and it stayed that way for another four decades. After the war, new routes to New York, Tokyo, Johannesburg, and Mexico City followed.

Sabena was one of the most noticeable airlines in Europe in the golden era of European aviation. In the post-WW2 period, Sabena was ranked seventh in route mileage, thanks to its European, North Atlantic, African networks, and domestic services in the Congo.

By 1960, Sabena was flying to 96 cities in 39 countries worldwide, with a fleet of 73 aircraft and five helicopters. It had 105 offices worldwide and employed 11,000 people. The airline was continually moving forward. On December 28, 1955, in the dawn of the Jet Era, it placed orders for three Boeing 707-329s. After the arrival of its first Boeing 707 Intercontinental on December 4, 1959, Sabena became the first airline in the world to provide a 707-300 jet service to the USA by a non-US-based airline when operations began on January 20, 1960.

There was a downside to the Belgian airline. Even though it was expanding rapidly and was improving its air traffic results yearly, there were problems with its finances. It was posting losses year in, year out. There were many reasons for the financial issues, such as high aviation fuel prices, currency instability, and persistent inflation. But, the catastrophe for Sabena was the downsizing of its primary route network to the Congo in 1960. The Congo declared independence that year, and Sabena was never the same again. Apart from the reduced Brussels-Kinshasa flights, Sabena lost the Congolese domestic route network, which accounted for 40% of the airline's overall revenue. That was enough to bring any airline to its knees. Sabena was also in dispute with the Congolese national carrier, Air Congo, and the Congolese Government due to the latter blocking Sabena's funds in a Congolese bank. Hence, Sabena could not use its funds in times of need, which was often.

Sabena soldiered on, and the traffic results kept improving while the financial issues kept ballooning. It was not a good situation for the national airline of Belgium. The Belgian State kept supporting Sabena financially, which did not please the Belgian people, as they were the ones keeping Sabena afloat with their taxes. Without that support, Sabena would have ceased to exist much earlier than 2001.

The story of this airline was not all bad. However, the most significant moment in Sabena's history was in 1960, when it stopped all its regular services and diverted its whole international fleet to the Congo to save the lives of the European and other residents in the country during the savage civil war. It was Sabena's most noteworthy moment. The airline's management, the flight crews, the flight attendants, the mechanics, the ground staff, and every other Sabena employee did their best to save the people from a hazardous situation.

Yes, Sabena did collapse in 2001, due to its many problems, just like many other airlines. Yes, Sabena was financially supported by the Belgian Government. But the history of this airline is as deep as they come. The modest nation of Belgium had an airline that was a pioneer in air travel. In the 1930s, Sabena opened up Africa to the world. It was an airline that saw the world and said, "World here I come!" Sabena was not a large airline, but it was an airline that was regularly placed in the top 20 world airlines list during the golden years of civil aviation. Sabena indeed was a world airline.

ACKNOWLEDGEMENTS

This book would not have been complete without the photographs of so many talented aviation photographers. I want to thank the following photographers for their assistance in producing this book.

Richard Vandervord has an excellent eye for taking the right shot. One of the best photographers I have seen.
John F. Ciesla has some of the best classic airline photos I have ever seen.
Frank de Koster has many unbelievably outstanding photos taken at Rotterdam and other airports in Europe. I enjoy looking through his excellent website as well! Please check his 'The Mighty Eight' website at https://www.douglasdc8.com/
Eduard Marmet is also one of my favorite photographers. He has a gift for taking fantastic eye-catching photos.
Michael Bernhard is another photographer, like John Ciesla, who has taken some of the finest photos from the significant classic airliner era. A true Austrian master.
Stefan Röhrich. I want to thank Stefan for his marvelous photos and for providing me with the late Norbert Kröpfl's amazing airline images from the classic airliner era.
Udo K. Haafke provided many great photos from the 1970s to the 1990s.

I would also like to thank the following photographers for their contributions:
Bernhard Bauske, Nobuo Oyama, and Jeremy Hughes.

The Beginning
1919-1929

The first company in the world to use aircraft for revenue purposes was the German company 'Deutsche Luftschifffahrts-Aktiengesellschaft' (German Airship Travel Corporation) which began operations with airships in November 1909. After that, other airlines came into being. The second airline was the Hungarian 'Aero Rt.', which started operations in 1910. Following that, the first US airline to come onto the scene was 'St. Petersburg-Tampa Airboat Line' which was based in Florida and established in 1913. However, the introduction of commercial aviation in Belgium was a little overdue. It would be another six years before a Belgian Airline would make its appearance. Its origins lay in the Belgian Colony of the Congo in 1919.

Belgium's King Leopold II, with support from several Western countries, formed the Congo Free State on February 5, 1885, claiming the African landmass as his territory. The Congo Free State was a privately controlled colony by Leopold II, to the detriment of the local people, who were the victims of terrible abuses and murders under his rule. Many Europeans, who, at first, supported his takeover of the Congo, spoke out against the abuses in the 'Congo Free State' as the realities and suffering became more generally known. Subsequently, on November 15, 1908, the Belgian government annexed the 'Congo Free State' and took over its administration.

Following the Belgian State's acquisition of King Leopold's Congo Free State in 1908, the King continued to have a substantial role in Belgium's colonial affairs. Consequently, a 'Fonds Special,' or 'special fund,' was created, which consisted of a sum of money placed at the King's disposal as compensation for handing over his African holdings. He could use the funds to sponsor projects to improve the colony's economic and social well-being. King Leopold died just one year later, on December 17, 1909. His successor, King Albert I, would utilize the fund to develop aviation in the Belgian Congo. For that reason, he was given the well-deserved nickname of 'King aviator' by the Belgians.

On November 12, 1918, the day after the armistice of World War I was signed, King Albert I set up a committee to study the possibilities of commercial aviation in the Belgian Congo. The committee saw great opportunities for civil aviation in the vast, undeveloped colonial territory of the Belgian Congo, which was approximately 80 times the size of Belgium. On June 26, 1919, CENAC, the 'Comité d'Etude pour la Navigation Aerienne au Congo' (Study Committee for Air Navigation in Congo), was established by a Royal Decree and financed by the 'Fonds Special.' CENAC's role was to experiment with creating an airline for passengers and postal transport in the Belgian colony of the Congo. An appropriation of BFR 2 million was drawn from the 'Fonds Special' and was made available to the committee.

On January 15, 1919, a young officer in the Belgian military, Georges Nelis, released a booklet pushing for establishing a Belgian national civil aviation company. He was not the only visionary thinking of a national airline for his country. The need for a national airline had been in the minds of many influential Belgian bankers for quite a while. Ultimately, the banking community is credited for forming Belgium's first airline. The new airline was founded on 25 March 1919 and named the 'Syndicat National pour l'Etude des Transports Aeriens' (The National Union for the Study of Air Transport). That name did not last long, as in November 1919, it was renamed 'Societe Nationale pour l'Etude des Transports Aeriens', 'SNETA'. The only change was the 'Syndicat' being replaced by 'Societe.' The bankers decided to name Georges Nelis, a Belgian Air Corps officer, SNETA's director-general. Thus,

Belgian air transport had SNETA to thank for opening the horizon for a future of air travel for the Belgian nation. In 1920, SNETA began operations with flights from Brussels to London and Paris.

CENAC entrusted SNETA to carry out the possibilities of creating an airline in the Belgian Congo. Then, in early 1920, the two organizations formed an airline and named it in honor of King Albert I, who was the driving force behind the airline project. The new airline was LARA ('Ligne Aerienne Roi Albert' - King Albert Air Line). LARA was the first colonial airline to operate south of the equator and one of the world's first regular air transport companies. Testing started in July 1920, when the first section of the route from Leopoldville to N'Gome 360 miles (580 km) was put into operation. Afterward, LARA inaugurated the following routes in the colony, July 10, 1920, Leopoldville-Coquilhatville 491 miles (790 km), March 3, 1921, Leopoldville-Lisala 739 miles (1,190 km), and July 1, 1921, Leopoldville-Stanleyville 1,072 miles (1,725 km). The latter route was the most important one, as it serviced the two biggest cities in the Belgian Congo. At the end of the experimental flight operations, the results of the test flights were as follows: LARA completed 1,000 flight hours, covered more than 77,671 miles (125,000 km), and made 80 connecting flights. In June 1922, after a year of conclusive tests, LARA suspended its operations due to high costs and closed down in June 1922. As a result of the tests, SNETA and the Belgian Government were convinced of the feasibility of establishing regular routes.

Photo 1. King Albert I played a significant role in developing civil air transport in Belgium, and was nicknamed 'King Aviator' by the Belgian public.
(Photo via Wikimedia)

Sabena is born

1923

The Belgian Government and SNETA made preparations to create a new airline that would replace LARA and represent the Belgian nation. Consequently, the Statute of Sabena was submitted to the Belgian legislative chambers on April 23, 1923. The new airline was legally incorporated and officially announced to the public on May 23, 1923. Its name was 'Societe Anonyme Belge d'Exploitation de la Navigation Aerienne,' SABENA. Sabena's share capital was set at BFR 6 million, consisting of 12,000 preferred shares at BFR 500 francs. SNETA participated in Sabena's formation by subscribing 50% of the authorized share capital, with the Belgian State and the Belgian Congo subscribing to the remaining 50%. SNETA held 6,000 shares, the Belgian State 5,800 shares, and the Belgian Congo 200 shares.

Although Sabena was 50% owned by the Belgian Government but was run like a private enterprise. The airline was governed by a Council of 12 members chosen for a six-year term by the General Assembly of stockholders. Two were selected from candidates provided by the Minister of Civil Aviation, one from candidates proposed by the Minister of the Colonies, and one from candidates proposed by the Minister of Finance. The Government had complete influence over Sabena's policies owing to these four individuals. They possessed veto power over the Council's decisions. The assembly elected the remaining eight members after receiving nominations from stockholders other than the Belgian Government and the Belgian Congo.

Even though the airline was run like a private operation, it had the firm support of the Belgian Government. For instance, the government and the Belgian Congo pledged a five-year subsidy to cover the Belgian carrier's deficit up to a fixed sum. In addition, Sabena was granted lifelong concessions by the Governments of the two states. Sabena was assisted with the introduction of the statutes which would make the airline bulletproof in the face of financial problems. The announced regulations were as follows: 1) The Belgian State granted Sabena a right of preference for the transport by air of mail and parcels in Belgian territory, both on domestic and international flights. 2) The Belgian Congo granted Sabena the exclusive concession for the government transport by aircraft, mail, parcels, passengers, and goods on all routes operated by Sabena in the territories of the Belgian Congo, both on domestic and international flights. 3) For the purchase of airliners needed for operations in Europe and the Belgian Congo, the Belgian Minister of Finance for Europe, and the Belgian Minister of the Colonies, could issue bonds, which the Belgian State and the Belgian Congo would guarantee. 4) For operations in the Belgian Congo, as a priority, Sabena would provide flights for the Belgian Congo Government, and the various administrations of the Colony, up to 80% of the capacity available for each regular flight. Moreover, Sabena would be required to make a minimal annual number of such journeys, and the distribution during the year would be agreed upon between the Belgian Congo and Sabena. This statute allowed the privilege of free flights for Belgian officials. It also meant that the Belgium-Congo route would guarantee the Belgian carrier regular revenue, calculated according to the fares in force at the time. 5) For the Belgium-Congo route, the Belgian Minister of Transport and the Belgian Minister of the Colonies would conjointly determine the maximum amount of the subsidy to be granted to Sabena to cover any operating deficit on the Belgium-Congo route. If there were any excellent years, the net proceeds would be equally shared between Belgium and the Belgian-Congo. Unfortunately, there would not be many good years.

Photo 2. A Handley Page W8b, 'O-BAHJ,' of SABENA's predecessor SNETA. (Photo by ETH Zurich)

1924

Sabena's first year of operations, in 1923, was confined to the carriage of newspapers on the Brussels-Ostend-Lympne line. However, things improved when on April 1, 1924, the Belgian national carrier began operating a route between Rotterdam, Brussels, and Strasbourg. The route was extended to include Basle on June 12. Initially, Sabena only transported parcels and general freight on the route, and on July 1, it started offering passenger services. On July 15, a dozen Swiss and Belgian journalists flew to Waalnaven, a harbor area at Rotterdam, on a Sabena three-engine Handley Page aircraft to celebrate the opening of the new passenger Rotterdam-Brussels-Strasbourg-Basel route. The Sabena aircraft they were on was built in Belgium under license from Handley Page, the large UK aircraft manufacturer. The aircraft could carry ten passengers, with 15 kg of baggage allowance per passenger. The flight left Basel at 8 am and touched down at Rotterdam at 2 pm, greeted and cheered by a large crowd. There were also many dignitaries there, including the consuls of Belgium and Switzerland, representatives from Sabena, and KLM Director, Albert Plesman. Later, on September 15, the line was further extended to include Amsterdam. Unfortunately, the route did not last long. On September 1, 1925, it was closed down by Sabena due to it being unprofitable. Later attempts to re-establish services were unsuccessful.

1925

The event that captured the imagination of the Belgian people in 1925 was the first flight from Brussels to Leopoldville. It was a significant accomplishment for Belgium to achieve the feat at the time. In those early years of airline flying, the transport of passengers outside of Europe was seen by many as a dangerous adventure. However, in 1925, one man threw caution to the wind and proposed to open an air link from Brussels to Leopoldville. That man was Lieutenant Edmond Thieffry, the Belgian World War 1 air ace, aviation pioneer, and Sabena co-founder. Sabena went with the idea and offered Lieutenant Thieffry one of its newest planes. That aircraft was a Handley Page W8 which Thieffry named 'Princesse Marie-José,' after the youngest daughter of his friend King Albert I and who later became the last Queen of Italy. The introduction of the Handley Page W8 trimotor aircraft in 1925 significantly increased passenger safety and comfort, as it was considered a reliable and sturdy aircraft. The W8 type was Handley Page's first purpose-built civil airliner. On Thursday, February 12, 1925, Lieutenant Thieffry, pilot Leopold Roger, and mechanic Joseph Debruycker took off for the Belgian Congo. The distance covered was 5,048 miles (8,124 km), and the flight lasted 51 days, of which only 75 hours was pure flight time. That was due to the aircraft's slow speed at the time and the constant need to land for fuel. In the end, they succeeded in their endeavor. It was a watershed moment for Belgium, Sabena, Lieutenant Thieffry, and his crew.

Photo 3. Edmond Thieffry (left), with Jacques Lang and Philippe Quersin.
(Photo courtesy of Roland Michau. Photo by Joseph Lang)

Sabena successfully inaugurated the Leopoldville-Llebo-Lueba 528 miles (850 km) route in the Belgian Congo on April 25. M. Clere piloted the Handley Page W8 trimotor airliner. Interestingly, he was accompanied by Leopold Roger, the 'Princess Marie-José airliner pilot, who flew with Lieutenant Thieiffry, and Joseph Debruycker on the inaugural Brussels-Leopoldville line. The Sabena aircraft on the new route to Lueba carried just four passengers and 600 kg. of mail.

The performance of Sabena fin 1925, in both the European sector and the Congo, showed that the Belgian airline did well. In Europe, it covered 144,683 miles (232,844 km) and carried 4,403 passengers while moving 77,275 kg of freight and 10,967 kg of mail. Over in the Congo, the figures were much lower than the European sector but relatively good. Sabena flew 30,440 miles (48,989 km) and ferried 313 passengers, while its freight and mail figures were 77,275 kg and 10,967 kg, respectively.

1926

On February 9, 1926, Sabena opened a new route in the Congo, this time, and it was Luebo-N'Gule in the Katanga region in the southern Congo. The Belgian carrier thus extended its route network and offered the following routes: Leopoldville-Bandundu-Eolo-Pebeangu-Luebo, (350 km) and Luebo-Kanda Kanda-Kamina-N'Gulé 528 miles (850km).

During the first seven months of 1926, Sabena transported 221 passengers, and its mail service numbers increased dramatically. The airline's monthly mail carried in the Congo went up from 10 kg in January to 500 kg in July. In total, Sabena carried 21,725 kg of mail during those months.

The carrier's freight numbers were also impressive, as it transported 1,660 kg of goods in the same period.

Sabena's position improved in 1926 due to inaugurating new routes to Germany. The German route network was the beginning of what later became the central element in Sabena's European network. They were also the only routes in Europe that remained entirely Belgian in operation, as all the other routes were operated in pools. Sabena offered the following exclusively German routes: Brussels-Cologne-Berlin and Brussels-Cologne-Dusseldorf. The rest of the routes were Brussels-Cologne-Hamburg-Copenhagen, Brussels-Cologne-Frankfurt-Mannheim-Stuttgart-Zurich-Basle, Brussels-Amsterdam-Hanover-Berlin, Brussels-Amsterdam-Bremen-Hamburg-Copenhagen-Malmo, Brussels-London with two daily flights, and Brussels-Paris.

The new German routes did help the Belgian carrier, but the fact was that Sabena found it challenging to compete with the other larger European airlines for international traffic. It was handicapped by a lack of equipment and inadequate subsidies. Sabena did not have a strong bargaining position in making arrangements with its European competitors, as they had most of the air traffic in Europe. This situation would continue until the end of the 1920s.

1927

The Belgian Government granted Sabena a new 5-year subsidy contract. However, the agreement stipulated that the subsidy was paid separately by the Belgian State and the Belgian Congo. That was because the Government worked out two separate subsidy arrangements for Belgium and the Belgian Congo. The Belgian state's subsidy was based on the ton-km flown by Sabena and the expected deficit after that. In contrast, the Belgian Congo's subsidy guaranteed the equivalent to the revenue that an 80% load factor would produce at commercial rates. The following year, the duration of the subsidy contract was fixed at nine years, and the maximum assignable amount payable by the Belgian State and the Belgian colony during the period 1929-1937 was set at BFR 234 million.

In the Belgian Congo, Sabena, with its dependable three-engine Handley-Page W8 aircraft, provided air services to the following destinations, Boma-Matadi (the chief port of the Congo)-Leopoldville-Luebo-Elisabethville. In addition, the Belgian airline also introduced three new routes in 1927. They were the N'Gule-Elisabethville route 139 miles (225 km) opened on May 20, the Leopoldville-Boma route 217 miles (350 km), which began on December 14, and the Luebo-Tschikapa route, which was inaugurated on December 27.

The successes in the Belgian Congo were not through fate but by comprehensive planning and execution. From 1925-to 1926, Sabena tripled its total air mileage in the Congo and its ton-km amount. That was due to an elaborate program to enhance the carrier's air traffic in the Belgian Congo, which showed immediate results. Sabena, which was supplied with private and Belgian Government capital, focused on creating an air route network to connect all the essential areas of the Belgian colony. Cities, such as the capital Boma, the flourishing Katanga region, the mining center of Elisabethville, the Leopoldville, and many other smaller towns in the colony. Sabena wanted to ensure that all the parts of the territory would be reached in less than a day.

Looking at the numbers, during the first six months of 1927, Sabena carried more cargo compared to the whole year in 1926. More remarkable was that the 1927 total was achieved even though significant price increases were implemented in the colony at the start of the year. Sabena increased its ton-km from 49,989 in 1926 to an extraordinary total of 158,322 ton-km in 1927. Additionally, the 18,376 total for ton-km in 1926 almost tripled to 51,386 in 1927. Meanwhile, the total receipts, which amounted to BFR 920,149 in 1926, increased to an exceptional BFR 2,616,599 in 1927. Furthermore, in December 1927 alone, Sabena completed 93 flight hours, covered 6,935 miles (11,160 km), carried 80 passengers, and delivered 2,300 kg of mail.

1928

In 1928, a new domestic service was initiated between Brussels and Antwerp. However, the focus was on the Belgian Congo that year. On May 1, Sabena inaugurated a regular Luebo-Luluabourg-Lusambo service. The distance between those three cities was 201 miles (325 km) and was covered within one day. The new Luebo-Luluabourg-Lusambo route was subsequently linked to the Boma-Matadi-Leopoldville-Luebo-Elisabethville route, thus extending Sabena's network in the Belgian Congo. Later, on May 14, Sabena opened another Congolese route that connected Leopoldville and Coquilhatville 456 miles (735 km). The route flew via Bandundu and Mongo and was completed on the same day. In 1928, the main routes of Sabena in the Belgian Congo were Boma-Leopoldville, Leopoldville-Elizabethville, Leopoldville-Coquilhatville, Luebo-Lusambo, and Luebo-Chikapa). Flights were made every 15 days and were scheduled to make quick connections with the European ships at Boma, the capital and the main port of the Congo. Sabena's performance in the Belgian Congo, for the first six months of 1928, saw the carrier cover 67,137 miles (108,046 km), transport 879 passengers, and carry 11,641 kg of mail and baggage and cargo, all with efficient regularity. By the end of 1928, the carrier's network in the colony covered 2,485 miles (4,000 km), while its ton-km reached 130,151, and it transported 1,990 passengers for the year. Meanwhile, in Europe, the numbers were higher due to the vast route network Sabena had there. The carrier's aircraft flew 236,121 miles (380,000 km), transported 2,925 passengers, and carried 74,505 kg of freight.

1929

The third recapitalization of Sabena took place in 1929. The Belgian Government approved the capital increase of BFR 10 million to 20 million. Twenty thousand preferred shares were issued at BFR 500 each. That brought the total capitalization up to BFR 20 million. The Belgian Government held 5,000, SNETA 10,000, and the Belgian Congo 5,000. There were also 20,000 dividend shares issued without par value (The par value of stocks is the stated value at which they can be issued or redeemed. Stocks with no par value have no declared value, and their prices are set by how much investors are prepared to pay for them on the open market). The Belgian Government held 14,800 preferred and 23,800 dividend shares, SNETA held 20,000 preferred and 21,000 dividend shares, and the Belgian Congo held 5,200 preferred and 7,200 dividend shares.

In May 1929, Sabena inaugurated two new routes. They were the Antwerp-Dusseldorf-Essen-Hamburg and Brussels-Essen-Hamburg routes. In addition, Sabena's domestic route network offered flights to Antwerp, Liege, and Ostend from Brussels, and Liege and Ostend from Antwerp. From May 1 to September 31, Sabena transported 3,820 passengers on its international routes during the summer season. Meanwhile, on its domestic network, it carried 2,654 passengers. The total air traffic number for 1929 was 6,374 passengers, which was a big jump from the 3,000 passengers the previous year. Furthermore, Sabena transported 129,188 kg of freight and mail in 1929 against 112,207 kg in 1928.

The Pre-War Years
1930-1938

1930

On April 14, the Belgian national airline inaugurated a London-Brussels night mail service. A Fokker F.VIIb trimotor was designated for the new route. The first flight departed Brussels at 11.30 pm with 10,000 letters and telegrams and arrived at Croydon at 2 am. Sabena maintained several other routes in 1930. These were: London-Antwerp, Antwerp-Amsterdam, Antwerp-Liege, Brussels-Luxemburg, and Ostend-Liege.

In early 1930, Sabena ordered 4 Westland Wessex aircraft for the day-to-day operations of its routes and its mail service. The Wessex was a six-seater commercial cabin monoplane fitted with three Armstrong Siddeley Genet Majors engines. It provided accommodation for four people and had a top speed of 108 mph (174 kph). Sabena received the Wessex during the summer of 1930. Unfortunately, one of the four was destroyed in a hangar accident at Evere Airfield, in Belgium, in December 1934. Eventually, the three remaining examples, including the one in the photo, were purchased by Alan Cobham, a UK company, in March 1935.

Photo 4. Sabena's Westland Wessex OO-AGC. It was sold to Alan Cobham in the UK in March 1935.
(Photo by Flightglobal via Wikimedia)

1931

The main focus for Sabena in the early 1930s was the Belgian Congo. Plans were in progress to establish a new route connecting Belgium with its colony by the mid-1930s. That was a lengthy 7,500-mile (12,070 km) distance for any airliner in the pre-World War 2 era. The impetus for the new route to the colony was economic. From the early 1920s, there was a gradual increase in commercial activity between Belgium and the Belgian colony. The mineral wealth of the latter was supplying raw material for new processing and refining industries in Belgium. Additionally, the Belgian Congo became particularly important as a market for Belgian manufacturers. Moreover, the instability of international markets, which accompanied the worldwide depression of 1929, highlighted the importance of the Belgian Congo to Belgium's economy.

At that point, the Belgian Government began to consider speeding up travel time to the Belgian Congo. Consequently, the Belgian authorities entered into conversations with the French Government looking to establish an airline between Brussels and Leopoldville, to be operated by Sabena in conjunction with a company designated by the French Government. France and Belgium had an overriding interest in teaming up for the new route to the Belgian Congo. Initially, the two countries decided to open a new mail service to the Belgian Congo. They signed an agreement that stated that

Belgium would be tasked to fly the mail from the Belgian Congo to Europe and vice versa. It was no mean feat, as the mail from the colony in the mid-1930s represented almost 75% of all postal freight emanating from Central Africa.

Meanwhile, France was entrusted with establishing the infrastructure for the route across the Sahara and further into central Africa. The provisional agreement for establishing and operating an airline flying from Belgium to France and the Belgian Congo was signed in Paris on January 18, 1929. The two airlines signed the final agreement in Brussels on May 23, 1930, and, after some delay, it was ratified on May 13, 1931, by the two Parliaments. The final agreement allowed the French and the Belgian companies to collaborate on a Belgium-France-Congo postal air route.

After going over many different route options, the French authorities concluded that making the Belgium-Congo link via the Sahara desert was preferable to using the Nile route. Before arriving at the Belgian Congo, the route was to start from Paris and fly via Algeria, French West Africa, and French Equatorial Africa. Initially, the first section of the flight was from Paris to Libenge (a town in the far North West region of the Belgian Congo, bordering the Congo and the Central African Republic). It was the nearest town to Europe, and it bordered French Equatorial Africa. Afterward, that route would be replaced by Paris's newer route to Elisabethville. However, regular services would not start until February 1935. That year, the French Government set up a new airline, 'Regie Air Afrique,' to collaborate closely with Sabena.

On December 23, 1931, a Sabena test flight took off from Brussels, heading for Algeria. The Fokker VII-3m 'OO-AGI' nicknamed 'Leopold Roger' was used for the flight, which Prosper Cocquyt piloted. Apart from three crew, there was a particular passenger aboard the aircraft, Tony Orta, Sabena's director for Africa. The test flight was to find out where the flights in the future would be able to stop so that the infrastructure could be put in place. The stops for the forthcoming route were Algiers, Laghouat, Adrar, Reggane, Bechar, and Oran.

1932

On May 17, Sabena inaugurated a new European route, Antwerp-Dusseldorf-Berlin. A Fokker F.VIIb trimotor flew the route. Several aviation personalities at the airport in Antwerp witnessed the momentous event, including Belgian Air Force Major Van Crombrugge and Colonel Jules Smeyers, the Managing Director of SABCA, the Belgian national aircraft manufacturer, and Sabena. There were also representatives of the Prussian Ministry of Commerce, the Berlin City Council, and Lufthansa at the inauguration. In 1932, Sabena operated the following routes in Europe: London-Brussels-Cologne-Dusseldorf, Brussels-Dusseldorf-Essen-Dortmund-Berlin, Brussels-Essen-Hamburg-Copenhagen-Malmoe, Antwerp-Cologne-Dusseldorf-Essen, Le Zoute (a small municipality a few miles from the Belgian-Dutch border)-Ostend-London, Brussels-Rotterdam-Amsterdam, Brussels-Antwerp, and Antwerp-Dusseldorf-Berlin. During the winter, only the first two routes were operated daily, making 1,218 miles (1,960 km), compared to 3,449 miles (5,550 km) in summer.

Sabena's performance results for 1931 were released in the middle of the year. The Belgian carrier had 46 aircraft in its fleet that year. Twenty-nine of them were based in Europe, and they were as follows: 17 Fokker F.VIIb, 3 Handley Page W8F, 3 Handley Page W8B, 4 Westland Wessex IV, and 2 Fokker FII. The other 17 aircraft were operating in the Belgian Congo; they were the following types: 6 Fokker F.VIIb, 8 Handley-Page W8F, and 3 De Havilland 50. Sabena's route network stretched to 1,117 miles (1,798 km) in its European operations, and its aircraft flew 22,810 miles (36,710 km) per week while carrying 28,383 kg of cargo. Over in the Belgian Congo, its operations were not as significant compared to the European network with its numerous destinations. However, the Belgian Congo was a large state. Being 70% of the size of Europe, many of the flights in the Congo within the colony were extremely long. Thus, in the Belgian

Congo in 1931, Sabena's route mileage for the year was vastly more extended than the European one; it stood at 1,670 miles (2,688 km); however, its weekly mileage for scheduled flights was only 2,505 (4,030 km), about 90% smaller than the European network. The other reason was that, in 1932, Sabena operated two routes in the Belgian Congo, albeit quite lengthy; Boma-Leopoldville-Coquilhatville-Stanleyville, and Leopoldville-Port Francqui. On a positive note, Sabena's cargo services in the Congo carried 18,992 kg of freight in 1931. The relative success of the cargo sector was a good result for Sabena in the colony.

In 1932, the Belgian Government subsidized the Belgian carrier for BFR 17.3 million. In addition, it operated 32 aircraft on its European services, which was three more than in 1931, and 16 planes on its African route network, one aircraft less than in 1931. Most of the airliners in the Belgian Congo were Fokker F.VIIb or Handley Page W8F types.

1933

In mid-1933, Sabena was embroiled in a bribery scandal that tainted the airline's reputation in the eyes of the Belgian public. As a result, the charge of fraud and embezzlement was brought against four of the top Directors of Sabena. The council discussed the case at length, and the three accused were referred to the criminal court. The first was the famed General Van Crombrugghe, the former Director-General of Aviation at the Belgian Ministry of Transport. Sabena Administrator, Albert Marchai, was the second accused. Both men were charged with falsifying Sabena's balance sheets. The third member of the group Maurice Philippson was a banker at the well-known Brussel based Société Générale de Belgique, the largest bank in the nation, which later became a holding company, and controlled 85% of the heavy industry in Belgium. Philippson and Marchai were accused of securities law violations, namely the fictitious release of securities. A fourth member of the group, Colonel Jules Smeyers, the Managing Director of SABCA and Sabena, was saved from the court case, as he died a year earlier in Brussels, on August 2, 1932.

The scandal began back in 1927 and went on until 1931. As the Belgian Government heavily subsidized Sabena, the fraudsters found an opportunity to take advantage of the situation. Their manipulations consisted of making subsidiaries of Sabena, which were set up so that the subsidiaries were shown to make profits year in and year out. At the same time, all the losses were attributed to the subsidized Sabena, whose dividend of 6% was guaranteed by the State. The subsidiaries were SABCA, the aircraft manufacturer; SNETA, the aeronautical research company; and SABETA, an aerial photography company. When the fraud was in operation, Sabena lost almost half of its capital in four years and was close to going bankrupt. However, the fraud ring forged the balance sheets so that the liquidation of Sabena could be avoided to keep the fraud going.

Moreover, the State subsidies were allowed to purchase aircraft for Sabena. The tricksters took advantage of that situation by selling aircraft made by the subsidiary SABCA at exorbitant prices, most of which the crime ring took for themselves. From 1927 the total amount taken by the fraudsters was estimated to be at least BFR 6 million.

The Belgian carrier did have some good news in 1933. With its dependable, durable Fokker VIIb, and British-built Westland Wessex trimotor airliners in its fleet, Sabena never experienced a significant accident in its first nine years, including passenger and cargo services. That was a major achievement for the airline, as it was operating an extensive route network covering both Europe and Central Africa.

In the first eleven months of 1933, Sabena flew 620,534 miles (998,652 km), transported 12,850 passengers, and carried 28,255 kg of mail and 270,213 kg of freight and baggage. The Belgian carrier operated the services London-Ostend-Le Zoute and Brussels-Antwerp-Rotterdam-Amsterdam on its own. On the London-Brussels-Cologne-Dusseldorf route, it cooperated with

Imperial Airways of the UK. The other routes that it flew were Brussels-Essen, Mulheim-Hamburg-Copenhagen, and Mulheim-Dortmund-Berlin. Sabena flew those two routes in cooperation with Air France. In 1933, Sabena's European route network covered 1,439 miles (2,316 km). Sabena also operated a Brussels-Cologne night mail express service during the 1933 summer period.

1934

A Belgium-France-Congo air link had been on the minds of the French and Belgian authorities since the 1920s. Eventually, the two nations signed the Franco-Belgian convention on May 23, 1930. The agreement allowed the French and Belgian carriers to jointly organize the air services to the African continent. The flights would be once a week and provided alternately by the French airline and Sabena. The French authorities concluded that making the Belgium-Congo link via the Sahara was preferable rather than the longer Nile route.

Consequently, the French started establishing the infrastructure for the route over the Sahara in preparation for its opening. According to the agreement, the new service to the Belgian Congo would be weekly and be provided alternately by the French and Belgian airlines. For the French, they decided that the flights, which would stop over at Stanleyville, would be extended later to include Madagascar.

Although the agreement had been signed on May 23, it was not ratified until after a lengthy delay. That was due to the Belgian side, which expressed its dissatisfaction. It cited the significant expenditure and potential budgetary charges as reasons to postpone the 'Sahara' project's completion. In addition, it advocated using the British Imperial Airways route over the Nile, as many Belgian colonists used that route to fly to the Congo's eastern provinces and Katanga in the south. In a debate held by the Belgian Government, ministers were asked to support the efforts of those who wanted to establish an air service in the Belgian Congo that would connect Stanleyville to Elisabethville, with 'Imperial Airways' closest stopover points. However, the Imperial Airways route could only connect certain parts of the Belgian Congo, placing the colonists in other provinces in an unfavorable position. Moreover, the ministers stated that the Franco-Belgian route must end in Leopoldville, the capital of the Belgian colonial empire. As the French had intimated, they were not keen on continuing the flights to Madagascar.

By 1934, the disputes about which route was the preferred one subsided. Following a new examination of the issue, Belgian authorities finally declared themselves in favor of the French Belgium-France-Congo route. Sabena's management stated that it would be able to take on the expenses of the new route with the resources at its disposal and without requesting more money from the Belgian treasury. In addition, Sabena's board of directors decided to start a postal service to the Congo on October 1, 1934.

For the French, they were satisfied with the Belgian side, and Paris released a decree on May 16, 1934, authorizing the Minister of Air to begin test flights from Algiers to Brazzaville, the capital of the Republic of the Congo, which was just to the north of the Belgian Congo. The flights were on a temporary basis and took place during 1934 with large three-engine airliners. The route's infrastructure was virtually complete. In Mali, it ran from Gao to Zinder, in Niger, which was a distance of over 2,000 kilometers. The French West African Air Force had set up a series of relief fields and airfields that served as staging points and were well equipped with all the latest equipment. The road to the French and Belgian Congo was well and truly opened.

Sabena expanded its fleet when it placed orders for 3 Savoia-Marchetti S.73 trimotor airliners in October 1934. The S-73 was powered by Gnome-Rhone K-9 600 hp engines. It was a high-wing monoplane with accommodation for 18 passengers, three crew, and one flight attendant. It was of mixed construction, with wooden wings with three spars and a steel tube fuselage. It had fixed landing

Photo 5. Sabena Savoia Marchetti S.73P C-N 'OO-AGY'. The aircraft flew for Sabena from 1937 to 1940, when all air operations were suspended in Europe due to the war. (Public Domain photo)

gear and variable-pitch propellers. The planes were due to be delivered in 1935, with the first expected at Brussels on March 1, followed by the second on April 1, and the final one on April 20.

1935

The year 1935 was a groundbreaking one for Sabena. On February 23, chief pilot Prosper Cocquyt and his copilot Jean Schoonbroodt took off from Brussels in their Fokker VII-3m trimotor, 'OO-AGH,' named 'Edmond-Thieffry' for the Brussels-Leopoldville inaugural flight. The two pilots took the following route with the following stopovers: first-day Brussels-Marseille (France)-Oran (Algeria) second day, Bechar (Algeria)-Reggane (Algeria), the third day, Gao (Mali)-Niamey (Niger), fourth day, Zinder (Niger)-Fort-Lamy (modern-day 'N'Djamena, in Chad), fifth day, Fort Archambault (modern-day 'Sarh' in Chad)-Bangui (Central African Republic)-Coquilhatville (modern-day 'Mbandaka' in the Democratic Republic of Congo), and on the sixth, and final day they reached their destination of Leopoldville. It was a very long and tiring flight, with a distance of 5,188 miles (8,350 km) covered. The average mileage they completed daily was 932 miles (1,500 km), which amounted to 10 flying hours per day. The Fokker trimotor that the two pilots flew on was already outdated in 1935. It was simply too slow and cumbersome. Fortunately for Sabena, it would receive the 3 Savoia-Marchetti S.73 airliners in April. From then on, the flights from Belgium to the Belgian Congo would be completed in around five days, in stark contrast to the journey by boat, which took 30 to 35 days. From the route's commencement in 1935, there was continuous growth of air traffic on the line, which confirmed its value to Sabena. The decision to open-air links with its African colony was the right one.

After the inaugural flight in February, weekly services were maintained, and the route was flown by Sabena and its French partner 'Regie Air Afrique' on alternate weeks. 'Regie Air Afrique' was founded by French interests in 1925 to establish an air service from France to its island colony of Madagascar.

On March 25, the first of 3 Savoia-Marchetti S.73 trimotors arrived at Haren airport in Brussels, piloted by the chief pilot Cocquyt, who was the pilot on the inaugural Brussels-Leopoldville flight. The S.73's first scheduled flight was undertaken on March 28, with a flight to Croydon, in south London. The Brussels-London trip was completed in 1 hour 20 minutes, despite a headwind of 32 mph (50 kph). On the way back the next day, the airliner completed the 200-mile (325 km) route in 1-hour 15minutes. Then on April 1, the S.73 was put into service on the Brussels-Berlin line. Coincidentally, on that same day, a Junkers Ju-52 trimotor of Luft Hansa landed at Haren, inaugurating the Berlin-Brussels route.

On June 1, Sabena opened a new route originating in Brussels, stopping at Lille before terminating at Ostend. The Lille-Brussels-Ostend route proved to be quite successful. So successful that by September 15, just three and a half months after opening the route, Sabena had transported a total of 1,990 passengers. Then, at the end of 1935, Sabena's management went one step further and introduced a new Lille to London route to supplement the original Lille to Ostend line.

November 9, 1935, witnessed the inauguration of a regular air route to Africa, which was flown entirely by the French airline 'Regie Air Afrique.' The flight originated from Algiers (Algeria), with its destination being Antananarivo (Madagascar), with a stopover at Elisabethville in the Belgian Congo. The aircraft of choice was a French Bloch MB.120 trimotor. The aircraft, registered 'F-ANNX' and named 'Sirius,' was piloted by Charles Poulin. The two passengers on the flight were Hirschauer, the French Inspector of Air Services, and Tony Orta, the Director-General of Sabena. One week later, the route was flown from Marseille by Sabena, but the destination was Elisabethville for that flight. The flight was accomplished in a Fokker VII-3m, 'OO-AGI,' nicknamed 'Leopold Roger,' which was piloted by Jo Van Acker and Auguste Closset. On arrival at Elisabethville, the Sabena aircraft exchanged mail with the French Bloch MB.120, which had flown in from Madagascar on the same day.

Photo 6. Sabena's famed chief pilot Prosper Cocquyt at Dübendorf Airport, Switzerland February 5, 1946. (Photo by ETH Zurich)

1936

In March, Sabena announced that the 3 Savoia Marchetti S.73 trimotors were due to be delivered in October. Then, in May, the Belgian carrier received two new Junkers Ju-52 aircraft from Germany, with registrations OO-AGU and OO-AGV.

On October 25, 1936, Sabena not only opened a new route but also broke a speed record. It was on the Brussels-Oran-Elisabethville 'Accelerated Air Service' route. The entry into service of the new three-engine Italian speedster, the Savoia-Marchetti S.73, made it all possible. With a little reorganizing of the schedules and adopting a more direct route to Stanleyville, the S.73 arrived at Elisabethville on October 29, just four days after departing from Brussels. Sabena made the right decision in purchasing the Italian stallions. At the end of 1936, on the Brussels-Congo route via Oran, Sabena transported 555 passengers, carried 11,459 kg of freight, and delivered 12,643 kg of mail.

1937

Nine years earlier, in 1928, Sabena agreed to a contract that specified Sabena would receive subsidies from the Belgian Government and the Belgian Congo. The period of the agreement was fixed at nine years. The maximum transferrable amount payable by the Belgian State and the Belgian colony during the period 1929-1937 was set at BFR 234 million. Consequently, it was revealed that of the total amount of BFR 234 million allowable in the subsidy contract, only BFR 160 million had been used by the carrier by the end of 1937, when the contract expired. It was a welcome sign that confirmed Sabena was on the right path. Consequently, the agreement was extended for another two years until December 31, 1939.

In October, Sabena was quite pleased with its Savoia-Marchetti S.73 trimotors. It ordered three more Savoia-Marchetti S.83 trimotors, which were due to be delivered in February 1938. The S.83 offered accommodation for ten passengers and was powered by 3 Alfa Romeo 126 R.C.34 9-cylinder air-cooled radial piston engines, each having 750 hp. Its cruising speed was 250 mph (400 kph) at 16,400 feet (5,000 m). The S.83 was similar in shape to the S.73, except it was 4 feet smaller but much faster than its predecessor. Its maximum speed was 276 mph, compared to the S.73s 205 mph.

1938

In the penultimate year before the Second World War, Sabena was preoccupied with enhancing its fleet. The Belgian carrier had previously ordered three new Junkers Ju-52S airliners, which were delivered in February and were placed on internal routes in the Congo. The number of Ju-52s increased to five after the delivery of two Ju-52s in 1936. The Ju-52s eventually replaced the three Fokker VIIs in the Belgian Congo. In April 1938, Sabena's total fleet in Europe and Africa consisted of the following aircraft: 7 Fokker F.VIIb, 5 Junkers Ju-52, 9 Savoia-Marchetti S.73, and 3 Savoia-Marchetti S.83.

In Africa, at the end of October, the frequency of the services to the Belgian Congo operated by Sabena and 'Regie Air Afrique' increased to two flights a week. At first, the schedule was operated with two departures each week by one airline, which later changed to one flight per week for each airline. The flights from Brussels to Leopoldville took four days and to Elisabethville 5 days. Subsequently, the French carrier, Regie Air Afrique, stopped at Brazzaville in the French Congo and did not go past that point. That meant that Sabena was the only airline flying to Stanleyville and Elisabethville each week.

In December 1938, on its African route network, Sabena transported 1,745 passengers, carried 20,030 kg of cargo, and delivered 16,726 kg of mail. That was achieved with just five pilots and five Fokker VII-3m trimotors.

The War Years

1939-1945

1939

On January 1, 1939, Sabena altered its Belgium to Belgian Congo route. It replaced Oran and Reggane with Algiers and El-Golea. Oran was situated on the shores of the Mediterranean, while Reggane was located in the western central part of Algeria. Algiers was also situated on the Mediterranean, but it was 255 miles further east of Oran, and El Golea (modern-day El Menia) was situated in central Algeria. It was a more direct route to the Belgian Congo.

Three months later, in April, Sabena divided the route from Belgium to the Congo in half. The first sector from Brussels to Gao (Mali) was flown by Sabena's Savoia-Marchetti S.83s, while the remaining part of the route, from Gao to Leopoldville, was flown by its Junkers Ju-52s.

In June 1939, Sabena released its performance figures for 1938. The Belgian carrier covered 2,012,403 miles (3,238,649 km), reached 17,980,748 passenger kilometers, and delivered 263,862 kg of mail. Its performance for 1938 was creditable.

Sabena's fleet in April 1939 was as follows: 5 Junkers Ju-52, 8 Savoia-Marchetti S.73, 3 Savoia-Marchetta S.83, and 2 Douglas DC-2, with 1 Savoia-Marchetti S.83 on order. In addition, its route network had expanded to cover the following routes Europe: Brussels-London, Brussels-Paris, Brussels-Cologne, Brussels-Ambers, Brussels-Dusseldorf-Essen-Berlin, Brussels-Frankfurt-Munich-Vienna, Brussels-Prague, London-Ostend-Le Zoute. Africa: Brussels-Marseilles-Algiers-Reggane-Belgian Congo, Leopoldville-Boma, Leopoldville-Lusando, and Leopoldville-Coquilhatville-Stanleyville.

In September 1939, after agreeing with 'Regic Air-Afrique,' Sabena increased its Belgium to Belgian Congo flights to twice per week. The flights were scheduled for every Sunday and Thursday. Unfortunately, the start of the war meant that all flights to the Belgian Congo were suspended. After the traumatic First World War, the dark war years had come to Belgium again.

1940

The Belgian Congo flights were partially restored in February 1940, with Marseilles serving as the origin due to the war. There was a departure from the southern French city every Sunday, with the aircraft arriving at Leopoldville and Stanleyville on Wednesday, followed by Elisabethville on Thursday. That was the only service Sabena was offering in early 1940, which was based on the route provided in its prewar July timetable of 1939. Belgian passengers who wanted to go to the Belgian Congo traveled by train to Marseilles to catch the flight. Regrettably, only three months later, with the invasion of Belgium in May 1940, all of Sabena's European operations were suspended. Sabena moved its head office to Algiers, where it started to assemble as much of its European fleet as possible. Then, when the French-Italian armistice was signed on 25 June 1940, the French Vichy authorities stopped the planes from flying to the French colony of Algeria and handed them over to the Italians.

Things got worse after the French surrender on July 25. The Belgian planes, which were already in Algiers and Oran, were placed at the disposal of the Italian air force by order of the Franco-Italian armistice commission. Sabena lost eight planes in the process, three Savoia-Marchetti S.83s, four Savoia-Marchetti S.73s, and a solitary DC 3. The worst part was that Sabena had 40 employees in Algeria, most of who returned to Belgium. However, a few of them decided to stay in Algeria, while others fled to the UK. It was a disaster for Sabena. The war had interrupted all of Sabena's operations in Europe. Consequently, when the Belgium-Congo connection was severed, flights and shipping services between Belgium and the colony were similarly suspended. Commercial activity in the Belgian Congo was paralyzed, resulting in Sabena's operations in Africa being put off, but not for long. Tony Orta, Head of Sabena's African operations, was joined in the Congo by Sabena's President Firmin Van Bree and Managing Director Gilbert Perier. Belgian civil aviation established itself in Africa with seven Fokker VII3m trimotors and 6 Junkers Ju-52s.

Undaunted by equipment loss and wartime disruption, Sabena quickly reestablished operations independently and collaborated with the British carrier, BOAC. In August 1940, Sabena resumed flights on the Belgian Congo's domestic routes. Later, Sabena's fleet was supplemented by 9 Lockheed L-18 Lodestar aircraft, seven of which Sabena purchased from the USA and two of which were offered to Sabena by the Congolese Government. BOAC also contributed by giving Sabena two Junkers Ju-52s. In the meantime, at the request of the British Government, regular services were also operated by BOAC on the Takoradi (Ghana) to Cairo route via Accra (Ghana), Lagos (Nigeria), Douala (Cameroon), Bangui (Central African Republic), Libenge (Belgian Congo), Stanleyville, Entebbe (Uganda), Juba (Sudan), Malakal (Sudan), Khartoum (Sudan), and Wadi Halfa (Sudan). The route was beneficial to the people living in the Belgian Congo, as Sabena's international network was not what it used to be. The BOAC routes stayed in operation until the end of 1945.

On September 27, 1940, two French 'Regie Air Afrique' Lockheed 14s, F-ARIY, and F-ARIV headed for France were seized during a refueling stop at Elisabethville and handed over to Sabena after getting re-registered. Naturally, Sabena made good use of them.

1941 - 1945

In June 1941, Sabena inaugurated a new route between Leopoldville and Cape Town via Elizabethville and Johannesburg. After the war, Sabena changed the destination from Cape Town to Johannesburg. While providing regular services throughout the war, Sabena also undertook several special missions on the African continent and in the Middle East to support the colony's war effort.

On March 1, 1944, Sabena inaugurated a route from Leopoldville to Algiers via Lagos and Gao. Algiers was chosen as the destination because civilian aircraft were not allowed to fly over the Mediterranean sea during the war.

The company's network increased from 3,635 miles (5,850 km) in 1939 to 19,660 miles (31,650 km) in 1944, which enabled Sabena to make a healthy profit. For four years, from Leopoldville, the Lockheed 18s, seized Lockheed 14s, Fokker VII-3ms, and the Junkers Ju-52s served the route network in Africa without the loss of life. Moreover, from 1940 to 1944, Sabena transported 34,543 passengers in Africa, including 9,190 in 1944 alone. The astounding yearly performances allowed Sabena to reach record occupancy rates of between 83% and 95% during those four years. Overall, the Belgian carrier achieved those feats, even though it operated the network with a chronic lack of aircraft parts. The spirit of Sabena's personnel enabled them to push themselves and their aircraft to unknown limits. Moreover, they managed to keep Sabena in flying condition with great ingenuity throughout the war. Remarkably, Sabena's staff from 1940 to 1944 consisted of only 86 Europeans and 441 native Congolese.

In September 1944, a service from the Belgian Congo to Britain was established. The service was via Lagos (Nigeria), Gao (Mali), Aoulef (Algeria), Casablanca (Morocco), Lisbon (Portugal), and terminated in London. The service could not be extended to Brussels until the Allied military authorities could give their approval, which they did on July 10, 1945.

Once it was back at its home base, Sabena went to work to re-establish its European network. By the end of 1945, its route network offered flights from Brussels to Paris, London, and Stockholm.

Meanwhile, on July 14, 1945, Sabena resumed the weekly Brussels-Leopoldville line with Lockheed 18 Lodestars. Later, on August 3, 1945, the last Belgian Congo to Britain flight took off from Leopoldville and arrived in Brussels on August 5. The line, which had been opened in September 1944, had served its purpose during the war and was terminated. Sabena survived the war with spirit and determination. Now, it was business as usual. The years ahead would be interesting for the Belgian carrier with the iconic curvy 'S' on the tail.

The Post War Years
1945-1949

1946

In 1946, Sabena further expanded its European route network. Firstly, in March 1946, Sabena, in collaboration with KLM, offered three daily flights between Brussels and Amsterdam. Soon after, Zurich was the next city in Sabena's network. In addition, Sabena added Geneva, Copenhagen, Oslo, Lisbon, Nice, Luxemburg, and Prague to the network by the end of the year.

In January, Sabena introduced a new airliner on the Brussels-Leopoldville line, the Douglas DC-3, OO-CBA. The aircraft had been delivered to the Belgian carrier a couple of weeks earlier, on December 26, 1945. Then on February 18, Sabena's newest airliner, a Douglas DC-4, OO-CBD, named 'City of Brussels,' arrived at Melsbroek Air Base in Brussels from the Douglas factory with a Belgian crew. Six days later, on February 24, Sabena inaugurated the new Brussels-Leopoldville route, via Algiers and Lagos, with the four-engine piston airliner. Three weekly flights were offered on the route. The first was on a DC-4, the second was a Lockheed 18, while the third, on Friday, was with the venerable DC-3. A second DC 4, OO-CBE, was delivered to Sabena just in time for the summer schedule, which enabled the carrier to offer four weekly flights from Brussels to Leopoldville. With the Douglas DC-4s, the flights to the Congo now only took 21 hours. The Douglas aircraft had a capacity of 49, 42 passengers and a crew of seven.

On May 20, Sabena took the DC-3s and Lockheed 18s off the Brussels to Leopoldville route. From that date, the Belgium-Congo line was exclusively flown by Sabena's DC-4s. In addition, the Belgian airline offered three weekly flights, which flew via Algiers and Lagos.

In April, Sabena announced that it would perform some test flights throughout 1946 to establish its first transatlantic route to New York. Three months later, on July 12, a Sabena DC-4 completed its first transatlantic round trip. The flight back to Brussels was completed in 16 hours and 50 minutes, with only one stopover made in Ghent. There were 36 passengers on board, including Colonel Fraiteur, the Belgian Minister of National Defense, and Leon Schoevalerst, commercial director of Sabena. The test flights were a success for the Belgian carrier. Sabena was gearing up for its first official flight to New York.

One month later, the Belgian company announced that it had put in a firm order, with the Douglas corporation, for three new Douglas DC-6 propliners, the delivery of which was due to take place in 1947.

1947

Sabena's route network reached 9,320,567 miles (15 million km) in 1946. That was an outstanding achievement compared to the 1,864,113 miles (3 million km) before the war. The Belgian carrier also made a profit of 2 million USD the previous year. On the European network, the number of passengers reached was 88,000. As for the Belgian Congo, the company operated an extensive network there in 1946 and transported almost 20,000 passengers.

Sabena's most important route in the European sector was the Paris-Brussels route, operated in a pool with Air France. At the time, the route was the busiest in the world. Sabena provided two daily flights, which increased to five daily flights in the summer of 1947. Since the war's end, the first flight from Brussels to Paris was on April 29, 1946. Air-France reopened the Paris-Brussels route with a DC-3, followed by Sabena with its DC-3.

Photo 7. Sabena DC-3, 'OO-AUN' at Dübendorf airport, Switzerland. c1946. In 1947 95% of all flights within Europe were undertaken by Douglas DC-3s. The aircraft was with Sabena from 1946 to 1951. (Photo by ETH Zurich, colorized)

Photo 8. Sabena inaugurated the Brussels-Zurich line on February 5, 1946. The aircraft which flew the route that day was this Douglas C-49D-DO (DC-3A), 'OO-CBC' It was with Sabena from 1946 to 1961. This photo was taken on February 5, 1946, after it arrived in Brussels. (Photo by ETH Zurich, colorized)

Sabena introduced other destinations in 1947 within Europe; Milan, Madrid, Lisbon, Deauville (France), Luxembourg, Dublin, Liege, Le Zoute, Basle, and between London and Le Zoute, via Ostend.

Sabena scheduled the following western European flights in September 1947 (all daily), 7 to London, 3 to Zurich, 3 to Amsterdam, 1 to Geneva, 5 to Paris, 2 to Luxembourg, and 1 to Basel (via Luxembourg), and 1 to Nice. Then, from 15 June to 15 September, holiday flights operated daily on the following routes: Brussels-Le Zoute, 1 flight, Brussels-Deauville, 1 flight, and Le Zoute to London, two flights. In other parts of Europe, Sabena offered the following weekly services: 5 to Prague, 7 to Copenhagen via Hamburg, 2 to Stockholm via Hamburg and Copenhagen, 3 to Oslo, 2 to Dublin, 3 to Milan, and 3 to Madrid and Lisbon. It was a true feat of Sabena to re-establish and expand its European route network so quickly after the destruction of the war years.

On Sabena's all-important Belgium-Congo route in Africa, the carrier reached a level of efficiency and speed never seen before. To top it off, the frequency of the flights increased. From Monday to Saturday, Sabena offered six flights per week to the Congo on its Douglas DC-4s. One of those six flights continued to the wealthy region of Katanga, in the south of the country. The flights to the Congo took the following route, Brussels-Tunis-Kano (Nigeria)-Leopoldville. Later in 1947, Sabena opened a new route in Africa, but it was not in the Congo; it was its first direct route from western Europe to the inspiring city of the pharaohs, Cairo, with two weekly flights. The route was inaugurated on June 16.

Since Sabena was established in 1923, its planes traveled more than 26,718,961 miles (43 million km) or 1,086 times around the earth! They carried almost 400,000 passengers, of which just under a third were in 1946 alone. That represented a total of 195,731,925 passenger miles (315 million passenger-km).

Then, the big moment in 1947 regarding Sabena's chances of genuinely expanding its intercontinental route network was the delivery of its first of three Douglas DC-6 propliners in July. The Douglas DC-6 could fly where no other propliner could fly before. It was the game-changer for Sabena, as the aptly nicknamed 'Skymaster' would allow Sabena to expand its network faster than any time in its history.

Sabena's DC-6s had a larger fuselage, more powerful engines, and a higher performance than its predecessors. The DC-6 was powered by four Pratt and Whitney 2,400 hp (1,800 kW) R-2800-CA15 engines. Its cruising speed was 311 mph (500 kph), and it had a range of 4,584 miles (7,377 km). Its maximum takeoff weight was 52,567 lb. (23,844 kg). The standard seating capacity was 52 passengers, and for night flights, there were berths for 26 passengers. Sabena would go for the first option, and wisely so. In the summer, Sabena became the first airline in Europe to put into service the Douglas DC-6. Then two months later, Sabena's DC-6 made its first scheduled flight to New York, pushing aside the DC-4 in the process. The DC-4 had been flying the route to New York since June 4, 1947. The Skymaster was the master now.

Sabena released some interesting statistics in 1947, showing how much improvement the carrier had made since the end of the war. By the end of 1945, Sabena's 13 aircraft transported just 189 passengers. By October 1, 1947, less than two years later, the number of available seats had increased to 1,224 on 44 aircraft. In 1937, Sabena's traffic amounted to 8,982,308 passenger miles (14,455,644 passenger-km), in 1946, the total amounted to 95,303,710 passenger miles (153,376,455 passenger-km). The extent of the Belgian carrier's network in 1947 was 20,922 miles (33,670 km). Sabena's fleet at the end of the year consisted of the following types: 3 de Havilland Dove, 7 Lockheed Lodestar L-18, 17 Douglas DC-3, 8 Douglas DC-4, 3 Douglas DC-6, and 6 Convair CV-240.

1948

On January 1, Sabena established a weekly cargo service to the United States. By March, the cargo service to the US had increased by 100%. All the cargo aircraft were packed. That was a phenomenal achievement, considering the service was only in operation for two months. The spike was due to a reduction of US customs tariffs, which were also introduced in January 1948 by the US Customs service. However, the return cargo flights to Brussels flew back with less freight. That was due to the stricter Belgian import controls. In addition to the cargo service, Sabena also opened a speedy passenger service between Brussels and Johannesburg, with a direct connection from London, on May 26. The Belgian airline placed the Douglas DC-6s on the route. The total flight time for the one-way journey was 25 hours.

In May, the Douglas corporation recommended that all of Sabena's DC-6 fleet undergo technical modifications to improve the DC-6s performance and comfort. As a result, Sabena withdrew its DC-6s from the fleet for the duration of the repairs. After the changes were completed, the DC-6s were returned to Sabena in early June. The Belgian carrier put them back into service on the Brussels-New York route on June 15.

Also, in June, Sabena opened a new route to Bern, Switzerland on June 15. Flights to the scenic Swiss city were provided three times a week, on Tuesdays, Thursdays, and Saturdays, with departures from Brussels taking place in the early afternoon. Then, on July 10, Sabena opened a new line to Nice with daily flights.

The year 1948 was a special one for Sabena. It was the carrier's 25th Anniversary. Mr. Van Acker, Minister of Communications, Mr. Fratteur, Minister of National Defense, and Mr. Wigny, Minister for the Colonies, spoke to a large crowd at the Palais des Beaux-Arts in Brussels. Sabena's growth coincided with Belgium's economic growth in those first 25 years. The Airline carried 1,789 passengers in 1923; by 1948, the total was 186,127. Furthermore, a particular mark of esteem was offered to Sabena by IATA (International Air Transport Association). The Belgian airline's president, Mr. Gilbert Perier, was elected president of that organization in 1948.

In June 1948 alone, Sabena opened nearly 12,428 miles (20,000 km) of new routes. That was incredible compared to Sabena's pre-war network in Europe and the Belgian Congo, which comprised just 11,185 miles (18,000 km). Its comprehensive route network for 1948 was 34,175 miles (55,000 km). The extension of the network was made possible by the delivery of newer, bigger, and faster airliners. Between December 31, 1946, and September 30, 1947, Sabena took delivery of many aircraft, and its overall fleet at the end of 1948 consisted of 3 Douglas DC-6, 7 Douglas DC-4, 1 Douglas C-54, 15 Douglas DC-3, 9 Douglas C-47, 5 Lockheed Lodestar, 3 de Havilland Dove. Small aircraft: 2 Piper Cub, 1 Auster.

Sabena served 20 major cities in Europe; London, Paris, Amsterdam, Oslo, Copenhagen, Stockholm, Hamburg, Basel, Geneva, Zurich, Nice, Milan, Rome, Athens, Lisbon, Dublin, and Prague. The Belgium-Congo route was crucial for Sabena and was now scheduled with daily flights to Leopoldville. For two times a week, the Sabena aircraft flew to Elisabethville, the rich mining region of Katanga in the south, and one of the two flights further extended the journey by flying to Johannesburg. There was also the important Brussels to New York route. Sabena placed its newest and fastest airliner on the route, the DC-6. In early 1948 Sabena was the first airline in Europe to place an order for the Convair CV-240, six aircraft in total, not forgetting that it was also the first airline in Europe to introduce the Douglas DC-6.

Photo 9. Sabena DC-4 'OO-CBH' just arrived at an airport in the Belgian Congo and was greeted by native dancers in November 1957. (Photo by ETH Zurich, colorized)

Photo 10. Sabena DC-4-1009 'OO-CBH' on the tarmac at an airport in the Belgian Congo in November 1957. The DC-4 was with Sabena from April 2, 1947, until August 9, 1963, when it was sold to Air Congo. The DC-4 was written off on April 7, 1974, in the colors of Air Zaire, when a downdraft caused it to land 100 meters short of the runway. (Photo by ETH Zurich, colorized)

1949

On November 29, Sabena announced that one of its DC-6 propliners, piloted by Captain Jean Moreau, broke the Brussels to Johannesburg route's speed record by flying to the South African city in just 20 hours and 40 minutes. There were 48 passengers aboard the flight, who must have been quite pleased to get there in such a quick time.

On December 15, 1948, Sabena inaugurated an air cargo service between Brussels and Leopoldville. It was the first regular cargo service carried out by Sabena to Africa and was also the first cargo service connecting Europe with Africa. The service was provided by the twin-engine Douglas C-47s, which offered a payload of 3 tons, and the four-engine Douglas C-54s, which offered a payload of 8 tons, almost tripling the C-47s amount. The new service was initially flown twice a month, with the cargo planes leaving Brussels on the 1st and 15th of each month. The new cargo service was made available due to Sabena's inspiration from the US situation, where all-cargo services only existed and operated in the US in the late 1940s. Sabena was one of the first airlines in Europe to follow the lead of the American cargo carriers.

In May, the first of six twin-engine CV-240 Convairliners ordered by Sabena arrived at Melsbrœck airport. Sabena speedily put it into service afterward. Sabena had placed the order for the aircraft at the end of 1947. Sabena's six new twin-engine CV-240 Convairliners had pressurized cabins, increased seating capacity for 40 passengers, and a speed of 298 mph (480 kph). Sabena mainly used the CV-240s from Brussels to London, Paris, Geneva, Zurich, Nice, Madrid, Lisbon, Milan, Rome, and Naples. The Convairliners considerably reduced flying time within Europe. For example, the Brussels-Milan flight took 2 hours, Nice 2 hours 15 minutes, and Rome 3 hours. Sabena intended to replace the DC-3s in its fleet with the Convairliners on all European routes.

On May 16, the Belgian carrier released its summer schedule. On the European network, Sabena offered the following routes with the new Convair 240 aircraft: Brussels to Paris, Geneva, Zurich, Milan, Rome, Naples, Madrid, and Lisbon. On July 18, the CV-240s replaced the DC-3s on the Brussels-Nice route. The Brussels to Paris route increased to three daily flights, two by Convair CV-240s and one by the DC-3. The Brussels-Basle route was extended to Berne three times a week with DC-3s.

Meanwhile, four daily flights to London were scheduled, all with CV-240s; however, the London-Antwerp route was still flown by the DC-3s. From June 1, three weekly flights connected Manchester to Le Zoute and Brussels, and on June 24, Sabena introduced a daily flight between London and the Belgian coast. From July 1, on the Brussels-Amsterdam route, Sabena and KLM offered discount tickets on return flights to promote tourism between the two Benelux countries. Three weekly DC-3 flights were provided on that route. In North America, the DC-6s were scheduled to fly three flights per week between Brussels and New York.

With a domestic Belgian-Congo network of 11,000 miles of routes in Central Africa, Sabena launched a service between Leopoldville and Inongo near Lake Leopold II in 1949. The new weekly service, which flew through Banningville (modern-day Bandundu) and Nioki, was flown by De Havilland Doves. In addition, the airline provided a new feature that allowed telegrams to be sent from the aircraft flying from Belgium to the Belgian Congo. Later, Sabena expanded the new service to local routes in the Congo region and Sabena's whole network, including the Brussels-New York route.

An Extraordinary General Meeting of Sabena shareholders was held on June 14, 1949, at the Company's registered office. As a result, by the law of April 6, 1949, and the King's decree of May 4, 1949, the Assembly unanimously decided to extend the duration of assistance to

Sabena for a new term of 30 years and increase the company's capital from BFR 20 million to BFR 300 million.

Then, in July, Sabena released its financial results. The carrier's capital increased by BFR 20 million to BFR 30 million, half of which has been jointly paid by the Belgian State and the Belgian Congo, and the other half by private shipping and colonial interests.

In October, Sabena released its financial results for 1948, which showed that the carrier sustained an operating loss of approximately BFR 18.9 million in the previous year. However, it was compensated by the contribution of an equal sum from a 1946 special provision. The company maintained a load factor of 62.5% and carried 146,111 passengers over 8,463,967 miles during the year. More passengers were transported the previous year, 186,127 in all. Moreover, Sabena's aircraft flew over 7,747,359 miles. Cargo and mail represented 25% of the company's traffic in 1948, compared with only 13% in 1947. The total number of Sabena employees in 1949 was 3,335.

In December, Sabena announced that Mr. Willen Deswarte was appointed President of the carrier. He took over from Mr. Tony Orta, who was promoted to airline counselor. Mr. Deswarte was previously the Managing Director's assistant since 1947.

Regarding the fleet, Sabena was pleased with its DC-6s and ordered two more DC-6 Superliners with air-conditioned cabins. The carrier already had 3 DC-6s in its fleet, which were used on the Brussels-Leopoldville-Johannesburg line and the Brussels-New York service. The two aircraft were due to be delivered by the following spring. As it stood at the end of 1949, the fleet was as follows: 3 DC-6, 7 DC-4, 6 Convair CV-240, 17 DC-3, 1 C-54, 7 C-47, 6 deHaviland Dove.

Photo 11. A Sabena de Havilland DH114 Heron 1B, 'OO-BIA,' seen here at Rotterdam The Hague Airport on January 10, 1968. This was the only Heron in Sabena's fleet. (Photo by Ken Fielding via Wikimedia)

The New Dawn
1950-1959

1950

Sabena launched a new 10,200-mile (16,415 km) route from New York to Johannesburg, with intermediate stops in Brussels, Rome, and Lydda at the dawn of the new decade. The service was allegedly one of the fastest globally, and it was initially operated every week with Sabena's 52-seater DC-6s. Then, on February 23, Sabena opened a line to Oslo for the first time since the end of the war.

Also, in February, Sabena put up for sale its fleet of de Havilland Doves, which were used on feeder services in the Belgian Congo. The DC-3s replaced them. Then, in July, Sabena acquired two more Douglas DC-6s, which brought the total to five DC-6 airliners in the fleet. The DC-6s enabled the Belgian airline to offer six weekly flights to New York, with services starting July 1. Flights were made available every day, except for Sunday. Also, on July 1, a second DC-6 flight was scheduled every Monday to Leopoldville, with the return flight leaving the African city on Tuesdays. On July 2, Sabena introduced flights to Athens and Tel Aviv with a DC-6 which flew via Milan once a week, every Sunday. Then, on July 3, a Sabena Douglas DC-6 was scheduled to fly the Brussels-Nice-Rome route every Monday.

Later, in July, Sabena announced an order for 2 Bell 47D, single-rotor, single-engine light helicopters manufactured by Bell helicopter company of the US. The Bell 47 D model was the first to appear with a molded 'soap bubble' canopy. The copter allowed for a crew of 2 and was powered by a 200-horsepower engine, giving it a top speed of 105 mph (169 kph). Bell set the delivery of the helicopters for August 1950. The Belgian carrier intended to use the copters for mail services within Belgium. On August 16, in conjunction with the Belgian Postal Administration, Sabena inaugurated the first helicopter postal service in Europe. The helicopters followed the following internal routes: Brussels, Libramont, Tongeren, Hasselt, Beringen, Turnhout, Herenthals, Antwerp, and back to Brussels. Departures for the helicopter mail runs started from Porte de Ninove at 11.45 am and ended at 3.45 pm.

On June 25, 1950, the North Korean Communist forces invaded South Korea, starting the Korean War. The United Nations held an emergency Security Council meeting on June 27, passing a resolution offering assistance in the evacuation of all foreign forces operating in Korea under the authority of the UN. In the large-scale airlift, named 'Operation Pacific' by the US military, several allied Air Forces and some airlines, primarily American and some from allied countries, offered their assistance in the evacuation efforts. During the operation, thousands of airline employees from the US went to Korea, working 24 hours a day, 7 days a week on a round-the-clock basis. Scheduled air carriers participating in the 'Pacific Airlift' operations flew 110,000 passengers and 14,800 tons of cargo to and from Japan until 1951.

Sabena played a small but essential part in the airlift when on August 5, its first DC-4s took off for Korea to assist in the airlift with two crews, followed by a second Sabena DC-4 on September 4, also with two crews. On October 15, the third and final, Sabena DC-4, took off for the Asian nation. The three Sabena DC 4's, which were offered to the UN by the Belgian government for the Pacific Airlift, completed their 100th San Francisco-Japan flight in March 1951. That was just after an eight-month presence in Korea. The Belgian DC-4s also flew 1,172,341 miles (1886700 km) and covered nearly six million ton-km, or 2% of the total ton-km produced on the Pacific Airlift, up to that time. The End of Sabena's participation in the 'Pacific Airlift' came on August 15th, 1953, when the two DC-4s, lent to the UN for two years by the Belgian Government, flew back to Belgium. They had made 194 San Francisco to Japan return flights when they were in the region. The airlift in Korea would not be the only one for Sabena; in 1960, it would take part in a much larger airlift operation closer to home.

One significant event for Sabena in 1950 was the complete relocation to its new home base, Melsbroeck, national airport in March. Sabena was the last airline to leave its old headquarters at Haren airport, a former military airfield and civil airport in Brussels. Moreover, Sabena was the only airline still using Haren airport for its DC-3 flights. As all regular flights of Belgian and foreign airlines took place from Melsbroeck as of February 1949, Sabena had to move to the new airport.

Finally, in July, Sabena announced that in 1949 it completed 12,686 flights, of which 2,650 were on the Belgian Congo network and 247 on the North Atlantic route network. The rest were in Europe and the Near East.

1951

In April, Sabena's Board of Directors confirmed the order for six new Douglas DC-6Bs from the Douglas Corporation. The propliners were scheduled to arrive at Brussels in March and April 1953. The more contemporary Douglas DC-6B was a big step up from the original Douglas DC-6. Sabena used the older DC-6s on its trans-Atlantic route network, flights between Belgium and the Belgian Congo, and services to the Middle East. The DC-6B had more power, greater cargo capacity, and a longer fuselage than the standard DC-6, which meant more passenger seats. It was powered by four 1864-kW (2,500-hp.) Pratt & Whitney R-2800-CB17 Double Wasp 18-cylinder, air-cooled radial engines, giving the DC-6B a cruising speed of 315 mph (507 kph). That was slightly faster than the standard DC-6's cruising speed of 311 mph (501 kph). However, the DC-6B's trump card over the older model was its passenger accommodation numbers. The DC-6Bs could carry between 42 to 52 passengers in a standard configuration or up to 88 in a high-density arrangement. On the other hand, the DC-6 could only offer accommodation for 48 to 56 passengers, with up to 68 in high density. At the beginning of 1951, Sabena's main fleet consisted of the following aircraft: 5 Douglas DC-6, 7 DC-4, 26 DC-3, and 6 CV-240 Convair Liners.

Sabena's performance figures, released in May, showed that the carrier achieved increases across the board. It transported a record total of 195,466 passengers in 1950, 15% more than in 1949. That was not all; the most spectacular increases occurred in the cargo and mail divisions, which rose dramatically to 72.4%, representing 28.3% of the airline's total business. Its total ton-km amounted to 38,673,195, a jump of 27.1% over the previous year's total. Furthermore, compared with 1949, the number of flights offered by Sabena increased by 24.4%, revenue rose by 31.8%, while expenditure showed only a 19.7% increase. The load factor on the regular routes had improved, and the decrease in costs led to a sharp rise in the volume of flights. Sabena performed just as well on the financial side, if not better. In the 1950 fiscal year, its operating and miscellaneous revenues amounted to BFR 925.2 million, increasing by more than BFR 220 million or 31.8% over 1949's amount. All depreciation and amortization expenditures and expenses totaled BFR 937 million, up 19.7% from 1950. The Company's revenues covered 98.7% of the costs in 1950, whereas revenues only covered 88.8% of the previous year's expenses. In compliance with the Company's statutes, this loss was compensated by a recoverable intervention of the Belgian State in the amount of BFR 20.6 million. The state had little to complain about in 1950, as in 1949, it was obliged to compensate Sabena more than four times the amount - BFR 89.1 million. It was one of those years that Sabena could do little wrong.

Photo 12. Sabena DC-6 'OD-ACY' at Heathrow in 1960. It was delivered to Sabena on September 9, 1947. (Photo by Jerry Hughes, colorized).

Photo 13. Sabena DC-6, 'OO-SDA,' on the tarmac at Geneva-Cointrin airport in March 1953. The DC-6 was delivered to Sabena on July 31, 1951, and sold to Air France on December 12, 1956. (Photo by ETH Zurich, colorized).

Photo 14. Sabena DC-6B 'OO-CTM' at Liverpool airport on November 25, 1964. It was delivered to Sabena on January 13, 1954, and sold to the West German Air Force in June 1965. (Photo by Ken Fielding via Wikimedia).

Photo 15. Sabena DC-6 'OO-SDC', seen here at Zurich airport in 1957. It was delivered to Sabena on September 9, 1947. It is the same aircraft as in photo 12. (Photo by ETH Zurich, colorized).

The number of passengers carried increased from 168,604 in 1949 to 195,466 in 1950, while the average distance traveled remained above 1,300 km per passenger, which increased the ton-km volume from 234 million to 270 million. The cargo division reached a quarter of the total volume transported in 1950, compared with 20% in the previous year, and mail transport continued to grow, from 1.27 ton-km in 1949 to 1.66 ton-km in 1950, Sabena was in good stead.

Over in the Belgian Congo, Sabena in Africa also performed well in 1950. The company's ton-km offered increased from 16.4 million ton-km in 1949 to 22 million ton-km in 1950, rising by 34%. Meanwhile, the ton-km used went up from 1 million ton-km to 16.6 million ton-km, an increase of 51% over the previous year. On the African domestic network, the ton-km offered increased from 5.8 million ton-km in 1949 to 6.3 million ton-km in 1950, an increase of 8.5%. Likewise, the ton-km used rose from 3.8 million ton-km to 4.3 million ton-km. In 1950, Sabena in the Belgian Congo carried 21,840 passengers between Belgium and the Congo and 29,270 passengers on the Congolese domestic network. That was an impressive amount if we see that in 1950 the total European population in the Belgian Congo, and surrounding countries, was about 56,000. On December 31, 1950, Sabena's personnel consisted of 4,804 persons, including 272 aircrews, 1,509 office workers, 1,447 general staff, and 856 indigenous Congolese. Overall, there was a staff increase of 18.9% in 1950.

August 21 was the first anniversary of Sabena's postal helicopter service, which it operated on behalf of the Belgian Postal Service. It was a first for Belgium and the first of its kind in Europe. Sabena operated the service on weekdays only, over a 266 miles (428 km) long route, and served the following cities in Belgium: Brussels, Libramont, Liege, Tongeren, Hasselt, Beringen, Turnhout, Herentals, Antwerp, and back to Brussels. The Mail was delivered and picked up at each city, while refueling was done twice during the deliveries. In the first year of operations, the postal helicopter service distributed more than 51 tons of mail, approximately 4.6 million letters, an average of 181.5 kg per day. The service was operated with the two Bell 47D helicopters. Sabena offered a very efficient postal delivery service, which attained a regularity of 91.2%. The whole operation from departure to arrival had to be completed in 4 hours and 42 minutes, which required an average speed of 54 mph (87 kph). Credit must be given to the helicopter pilots who did an exemplary job, considering they had to make 12 takeoffs and landings on the 266-mile-long (428 km) route, two of which were refueling stops. Three minutes were spent on the ground at each landing site and a little longer at the two gasoline stations. Sabena's helicopter mail deliveries were a massive success in their first year.

Regarding Sabena's route network in 1950, a new weekly Brussels-Leopoldville line was opened on September 26, using the Douglas DC-6s. The flights departed from Brussels every Wednesday evening and stopped over at Lisbon. That stopover meant that Lisbon received its first direct link between Portugal and the Belgian Congo, thanks to Sabena. We need to be reminded that Portugal had its own colonies in Africa, the most prominent being Mozambique and Angola. Thus the flight to the Congo would have been convenient for any Portuguese family or businessman flying to Mozambique or Angola, as they could transit via the Congo. After leaving Lisbon, the Sabena flight arrived at Leopoldville on Thursday at noon. The return flight left the capital of the Belgian Congo on Saturday evening and arrived in Brussels on Sunday morning. The route was a successful one. In August 1951 alone, 1,362 passengers flew to the Belgian Congo. That was a record number at the time. The other routes to the Congo offered by Sabena in 1951 were: Tripoli-Kano (in Nigeria)-Leopoldville-Stanleyville, and Brussels-Athens-Cairo-Stanleyville.

Sabena's winter schedule took effect on October 21. The number of connections between Belgium and Congo remained fixed at six. They were: Brussels-Rome-Kano-Leopoldville, Brussels-Lisbon-Kano-Leopoldville, two Brussels-Tripoli-Kano-Leopoldville flights, a Brussels-Athens-Cairo-Stanleyville service, and a Brussels-Tripoli-Kano-Stanleyville-Elisabethville route. The Brussels-Leopoldville connection via Rome was originally on Tuesdays but was changed to Mondays. The weekly Brussels-Stanleyville route via Athens and Cairo on Tuesdays was changed to Wednesdays from Brussels, returning from Stanleyville on Fridays. The Brussels-Leopoldville Thursday flight continued to fly on to Johannesburg.

On the North American, Brussels-New York transatlantic route, the frequency of flights, which the Douglas DC-6s operated, was reduced from 5 to 3 per week. In the southeastern Mediterranean area, the Brussels-Tel Aviv via Athens route remained unchanged, with the flight leaving Brussels on Monday and returning on Tuesday from Tel Aviv. Sabena opened a second line to Athens on Wednesdays from Brussels. The Brussels-Paris line received three extra flights on the northern European network, one on Monday, Wednesday, and Saturday. The reason for the increase was to offer direct connections to New York. As a result, the number of flights between Brussels and Paris increased to 30 per week. There were four daily flights offered between London and Brussels on the UK routes, including one via Antwerp, while on the Brussels-Manchester line, there were three weekly flights. The number of weekly flights between Brussels and Amsterdam increased to 32, and they were operated jointly by the Belgian, Dutch, and Scandinavian airlines. The three airlines also provided 14 weekly flights between Brussels and Copenhagen, five additional postal and cargo services, and seven flights between Brussels and Stockholm. In addition, the three carriers maintained joint operations with three weekly flights to Oslo and Kristiansand. Switzerland and Belgium were connected by six weekly Brussels to Zurich, and 5 Brussels-Geneva flights, operated by Sabena and Swissair. In Italy, Sabena offered two weekly flights with Convair CV-240s to Rome via Nice. Convairs were also used three times a week on the Brussels-Milan route. Sabena also offered flights to the Côte d'Azur (The French Riviera) with three connections, two of which flew at night. On the German and Austrian networks, all the flights were in collaboration with KLM, except for the Brussels-Hamburg line. These were the flights offered: Brussels to Frankfurt six, including two via Cologne, Cologne two, Dusseldorf five, Hamburg two, Munich two, Nuremberg two, and Salzburg two. The schedules for the Prague, Madrid, and Lisbon lines were left unchanged.

Photo 16. A Sabena Bell 47D mail copter is getting refueled at Rotterdam on May 18, 1953.
(Photo by Harry Pot (Anefo) via Nationaal Archief The Netherlands)

1952

In March, Sabena announced that it had broken a record in 1951 by transporting 260,674 passengers for the year, a big jump of 33.3%, from 1950. It was also the first time Sabena had surpassed the quarter of a million mark for passenger numbers.

In the same month, the Belgian carrier placed a new order for another 2 DC-6Bs, bringing its total orders to eight DC-6Bs, with the six already ordered from the Douglas Corporation. The first six four-engine Douglas propliners were due to be delivered in 1953, with the final 2 in early 1954. By that time, Sabena's fleet of the 'sixers' in the fleet would amount to 18. The orders of the 8 DC-6Bs were due to Sabena's management's desire to renew the airline's DC-6 fleet by replacing its older DC-6s with the newer and bigger DC-6Bs.

Further in March, Sabena stated it would offer new tourist-class services on the Brussels-New York route starting May 1. The DC-6s would fly the route three times a week in combined standard and tourist class accommodation. Tourist-class passengers would be accommodated in the main forward part of the DC-6s, with the standard-class passengers sitting at the aircraft's rear. That arrangement is quite different from our modern-day flight seating plans. On June 15, Sabena added a weekly DC-4 all-tourist class flight on the New York route. Then it scheduled four other tourist class flights, three flown by DC-4s and one with a DC-6 from July 1, which was in addition to the two-weekly standard-class DC-6 flights. The extra flights proved to be beneficial for Sabena. By the end of 1952, Sabena's transatlantic bookings rose to 13,000.

Back in February, the first air link agreement between Belgium and Spain was signed in Madrid on the 27th. Consequently, Sabena gained the rights to operate flights from Brussels to Madrid and Brussels to Barcelona, the two most prominent cities in Spain. In return, the Spanish national carrier would be able to fly the Madrid-Brussels, and Barcelona-Brussels routes. As a result, Sabena started flights to Spain on April 21, 1952, when a Sabena aircraft flew the Brussels-Barcelona route, with a connection to Las Palmas provided by a Spanish airline.

On the African network, Sabena became the first operator to introduce cut-rate fares on international routes to central Africa, which would open up the Congo to more tourists from Europe. Starting July 1, Sabena began operating nine weekly services from Brussels to Leopoldville, with one of those flights continuing to Johannesburg.

Regarding Sabena's routes, the President of the company, Gilbert Perier, stated in 1952 that Sabena's services were divided into four sectors. Their order of importance was: 1 Brussels-Congo, 2 the North Atlantic, 3 Europe to the Middle East, and 4 the Congo domestic network. Those networks represented 38%, 29%, 21%, and 12% of Sabena's activity in 1951, respectively.

In September, Sabena was unquestionably a devotee of the DC-6s and placed an order for two all-cargo Douglas DC-6As, which had a cargo capacity of 11 tons. Delivery was slated for the summer of 1954. As mentioned, Sabena had ordered 8 DC-6B propliners, and now, with the two DC-6A all cargo liners, the order book stretched to 10 aircraft. The new order for the cargo aircraft reflected the desire of the Belgian carrier to continue modernizing its fleet, keeping pace with technology, and developing Sabena further by adding new routes to its network. In addition, there was the cargo division. Sabena was among the top airlines globally that provided regular cargo services. The Belgian carrier's percentage of cargo transportation, in relation to its overall traffic, was the highest in the world, with 36.8% of its business in 1951. Not bad for an airline from the small nation of Belgium. The company undertook a detailed study of its cargo traffic for 1951. The results revealed that increases in its cargo services had been across the board in all 73 cities. Thus, the need for more cargo capacity.

Sabena's fleet at the end of 1952 consisted of the following aircraft: 5 Douglas DC-6, 7 DC-4, 6 Convair CV-240, 25 DC-3, 4 Dove, and 3 Bell 47 helicopters. It had 8 DC-6B and 2 DC-6A airliners on order.

Photo 17. Convair CV240-12 'OO-AWP' of Sabena at Zurich airport on August 9, 1956. It was delivered to Sabena on March 19, 1949. It was sold to LOT of Poland on October 7, 1957. (Photo by ETH Zurich, colorized)

1953

Sabena's first of eight DC-6Bs arrived at Melsbroek airport in Brussels on March 10. The pilot chosen to fly the very long 5,950-mile (9,576 km) delivery flight from Santa Monica was Captain A. Vernieuwe. The aircraft made only one refueling stop at Gander. Many VIPs greeted him on landing in Brussels, such as the Belgian Transport Minister Mr. Segers, the airline's president, Gilbert Perier, and Mr. Cowen, the US Ambassador to Brussels. Later in March, when more DC-6Bs were delivered, Sabena started placing the new DC-6Bs on its regular transatlantic services to New York to supplement the five older DC-6s flying the route. The addition of the DC-6Bs enabled Sabena to increase the number of seats available on the New York line by 75%. The additional DC-6Bs to be delivered were placed on Sabena's various European and Middle Eastern routes. Sabena also used them on the crucial African routes to the Belgian Congo and South Africa. The DC-6Bs were all configured to accommodate 77 passengers and were used solely for low-cost flights. Sabena's increased passenger capacity by adding the DC-6Bs was a substantial 33%.

The year 1953 was a defining moment for Sabena, as the airline started scheduled international passenger helicopter operations. It was the first airline in the world to open such a service. In April, Sabena's Board of Directors decided to purchase three Sikorsky S-55 helicopters for the new service. The S-55 helicopters had a payload of about 1 ton, a cruising speed of 93 mph (150 kph), and could carry seven passengers over 93 miles (150 km). The S-55 was powered by one Pratt & Whitney R-1340-57 9-cylinder air-cooled radial piston engine, offering 600 hp. The S-55 was heavily used by the US Army, US Marine Corps, and the USAF, with the military designation: H-19, and nicknamed 'Chickasaw.' Sabena's

S-55s were expected to be delivered at the end of the summer, at which time operations could begin. The company swiftly started talks to open international passenger and mail services originating from Brussels. Initially, Sabena planned to offer four routes: Brussels-Liege-Cologne, Brussels-Antwerp-Rotterdam, Brussels-Lille, and Brussels-Luxembourg-Saarbrucken. Sabena intended to open the following routes for the second stage: Brussels-Frankfurt and Brussels-Paris. It was a significant operation for Sabena, but it would put the airline at the forefront of innovative planning.

Sabena officially opened the first international helicopter service on August 3, when Captain A. Vernieuwe, the pilot who flew the first Sabena DC-6B to Brussels from Santa Monica, flew a Sikorsky S-55 from Brussels to Rotterdam. Following that noteworthy achievement, Sabena inaugurated other regular passenger services with its Sikorsky S-55 helicopters on September 1, using its Bell 47D and Sikorsky S-55 helicopters. The routes were: Brussels-Lille, Antwerp-Rotterdam, and Brussels-Liege-Maastricht. The S-55s conducted all flights. The airline also extended the Brussels-Liege service to Cologne and Bonn on October 5. In early 1954, Sabena planned to open the Brussels-Luxembourg-Saarbrucken line.

By November, the helicopter operations were not looking so good. Sabena cautioned that the operating costs of its helicopters were twice as high compared to the costs of airliner passenger traffic. The airline was beginning to have second thoughts. It claimed that the Sikorsky S-55 cost 1 US dollar per mile to operate. With only a capacity for the S-55s to carry seven passengers, the situation was problematic. Even with an occupancy of 77%, the helicopters were still making losses. But Sabena was not going to suspend helicopter operations. Sabena had received so many new passengers due to the popularity of the helicopters that it decided the helicopters were economically justifiable. In a press conference in November, Mr. A. Vernieuwe stated that after the first two months of helicopter operations by Sabena, it had become apparent to him that the helicopter was the only method of transport to take the place of the regular aircraft services in densely populated areas. He added that the helicopters transported 3,000 people during that period, of which 39% had never flown before, and 16% had continued their journey by regular airliner. Mr. Vernieiwe was not exactly right in saying the only way to travel in the future in 'built-up areas' was by helicopter. Much cheaper alternatives existed and still exist today, including the railway and subway networks.

On June 27, Sabena inaugurated a new route, Brussels-Luxembourg-Strasbourg, offering three flights a week with a Douglas DC-3. Later, on October 29, Sabena inaugurated the first air service between Manchester and New York. The Sabena DC-6B departed at 10.30 p.m. and landed at Idlewild Airport in New York 12 hours later. The Lord Mayor of Manchester, Mr. Alderman A. Moss, and other passengers were among those who were received by the Mayor of New York, Vincent R. Impellitteri, upon arrival at Idlewild airport. Captain Jaspis, Sabena's senior North Atlantic pilot, flew the aircraft. The interesting thing about the inaugural flight by Sabena, it caught BOAC's eye. Previously BOAC was not too concerned about opening a line from Manchester to New York, Although BOAC and the Manchester city authorities had been talking about the possibility of such a service for some time. After seeing the success of Sabena's action, the UK international airline changed its tune. It asked for permission from its civil aviation authorities to offer a weekly Boeing Stratocruiser flight from Manchester to New York. BOAC stated that the aircraft would only carry first-class passengers and depart from London, stopping over at Manchester, Prestwick, and Gander, before flying into New York.

Photo 18. Sikorsky S-55 'OO-SHA' at the Sabena Heliport at Maastricht, The Netherlands, on August 10, 1953. Delivered to Sabena on June 16, 1953, and sold to SNCASE, France, in 1955.
(Photo by Duinen (Anefo) via Nationaal Archief The Netherlands, colorized.)

1954

Sabena's flights on the Brussels-Liege-Maastricht-Brussels and Brussels-Liege-Bonn-Cologne-Brussels-Maastricht routes doubled to four in early January. That was a positive move, as Sabena had plans to expand its Western European helicopter network and increase its fleet of Sikorsky S-55s to seven. In the scheme, Maastricht would be an aviation gasoline stopover for its copters which operated the airline's mail delivery service on the following route: Brussels-Libramon-Liege-Hasselt-Tongeren-Beringen-Antwerp and back to Brussels. On January 18, Sabena started using its S-55 helicopters on its regular daily mail service.

The three Bell 47s were withdrawn from the mail runs and sold. The Bell 47s' mail route was 270 miles (435 km) long and was serviced by the helicopters from August 21, 1950, to January 16, 1954. The copters made over a thousand flights and carried 166 tons of mail, or about 33,236,000 letters, during their time in service. In addition, Sabena's Bell 47s provided the following daily flights with 95% regularity: Brussels-Libramont-Liege-Tongres-Hasselt-Beringen-Turnhout-Herentals-Antwerp-Brussels. Apart from the mail deliveries, the three Bell helicopters carried out many special missions, during which they transported 6,465 people. One of those missions was to The Netherlands to assist hundreds of people in distress by evacuating them from the heavily flooded areas during the catastrophic North Sea floods of 1953.

In late January, Sabena announced that it would replace its S-55 helicopters with Sikorsky's newest type, the S-56, by 1958. The S-56 could carry up to 40 passengers and featured a 72-foot 5-bladed fully articulated main rotor. Its tail rotor had four blades, and the helicopter was powered by a Pratt and Whitney Double Wasp R-2800-50, 1,900 hp, 18-cylinder, twin-row, air-cooled radial engine, offering a speed of 150 mph (241 kph). Sabena had hopes of using the S-56 on its Brussels-Paris and Brussels-London lines. However, although the S-56 was four times the size of the S-55 and could carry many more passengers, it was not a suitable helicopter for Sabena. Instead, it was a helicopter intended for the military, designed in response to a US Navy Specification for a Class HR (Assault) helicopter in 1950. In the end, Sabena purchased the Sikorsky S-58.

In March, the Belgian carrier revealed that in 1953 it carried 325,185 passengers, compared to 227,028 in 1952, a rise of 35%. Moreover, Sabena's aircraft flew 15,691,994 miles (25,253,817 km) in 1953, compared to 13,199,976 miles (21,243,303 km) in 1952. The increase was due to the introduction and expansion of the tourist class services and the introduction of larger airliners, such as the Douglas DC-6Bs. The volume of passenger traffic reached 514 million passenger-km, an increase of 36.3% over the corresponding figure for 1952. Sabena's cargo traffic reached 18.6 million tom-km, which rose by 25.3% in 1953. The load factor was 70.2% compared with 72.8% in 1952. The Belgian national carrier also pointed out that in the Belgian Congo, it carried 63,000 passengers the previous year, a notable figure since the non-indigenous population of the Belgian Congo did not exceed 80,000 inhabitants. At the end of 1953, Sabena's route network was over 62,000 miles (100,000 km) long and served 83 cities in 31 different countries on four continents. In its financial performance, Sabena made a loss of BFR 23 million in 1953.

The following month, Sabena's Sikorsky S-55 helicopters started operating from Brussels to Lille, Antwerp, Rotterdam, Liege, Maastricht, Cologne, and Bonn. Later, Sabena extended the network to include Arlon, Luxembourg, and Saarbrucken. Anselme Vernieuwe, Sabena's operations director, stated that the S-55s would soon start flying to the West German cities of Essen and Duisburg. He predicted that Sabena's helicopters would account for more than half of the ton-km flown on the airline's European routes by 1959. He also stated that Sabena's international helicopter services transported over 3,000 passengers the previous November. The Belgian airline's network comprised eight cities in four countries, serviced by the Belgian company's helicopters daily. Sabena's Sikorsky S-55s aroused great interest in the European community, as the carrier transported 1,600 passengers in September 1953 alone, which was the first whole month of the airline's helicopter operations.

Sabena's helicopter services reached the end of their first year in August 1954. Since August 1953, Sabena's helicopter services had made 3,500 scheduled flights and carried over 13,000 passengers. Its helicopters covered 548,049 miles (882,000 km) and delivered 26 tons of mail in their first eight months of operation in 1954.

In the Belgian Congo, Sabena offered the following international flights in 1954 from Europe to Africa and within Africa: Brussels-Leopoldville, five weekly flights (one via Lisbon, one via Rome, one via Geneva, and two via Tripoli). All five flights were with the DC-6Bs. Brussels-Leopoldville-Johannesburg, one weekly flight with a DC-6B. Brussels-Stanleyville, one weekly flight to Entebbe by DC-6B, then transfer to a DC-4 to fly the last leg to Leopoldville. Brussels Elisabethville two flights a week (one via Rome, Athens, Cairo, and Entebbe, and one via Athens, Cairo, and Entebbe) flown by DC-6Bs. Brussels-Bukavu (Belgian Congo)-Usumbura (the capital of Rwanda-Urundi) to Entebbe by DC-6B two flights a week, then transfer to a DC-3 to fly to the Congo. Johannesburg-Elisabethville, one flight a week with a DC-3. Entebbe-Stanleyville two flights a week with a DC-4. Nairobi-Entebbe-Usumbura-Bukavu, one flight a week with a DC 3. Dar es Salaam-Albertville-Usumbura-Bukavu one weekly flight.

Sabena's fleet in 1954 consisted of the following aircraft: 8 Douglas DC-6B, 5 DC-6, 7 DC-4, 23 DC-3, 4 Convair 240, 3 Sikorsky S-55, and some smaller aircraft. It had the following aircraft on order: 8 DC-6B and 1 DC-6A.

Photo 19. Sabena DC-7C 'OO-SFC' on finals at Manchester Airport in the UK, in 1962. The DC-7C was delivered to Sabena on December 20, 1956. It was later converted to a DC-7CF in 1961. Sold to Spantax in October 1967. (Photo by RuthAS via Wikimedia)

Photo 20. Sabena DC-7C 'OO-SFK' on the tarmac in Rome in 1958. Delivered to Sabena on April 28, 1958. Later sold to Spantax in February 1966. (Photo by Maurice Luyten via Wikimedia)

Sabena, which was the only European airline in 1955 providing international helicopter services, ordered two more Sikorsky S-55s. The new order brought the number of S-55s in service at Sabena to six. At the beginning of 1955, Sabena's helicopter services offered the following flights from Brussels: Lille, Antwerp, Rotterdam, Maastricht, Cologne, Bonn, and Liege. In February, the daily helicopter services on the Rotterdam-Antwerp-Brussels route expanded with an extra direct flight from Brussels to Rotterdam. In addition, there was a short stopover at Melsbroek on the return leg. Sabena's helicopter network expanded further in April with the addition of a new route connecting Brussels to Eindhoven, Duisburg, and Essen. As a result, there were three return trips on weekdays; one flight terminated in Duisburg, while the others continued to Essen. In 1955, the Sabena helicopter network encompassed the following cities, Antwerp, Bonn, Cologne, Liege, Lille, Maastricht, Dortmund, and Rotterdam.

Sabena was bold to start international helicopter services, and its international network was a unique endeavor. Initially, many airlines in Europe and beyond were studying the progress of Sabena on its services, and they were somewhat impressed. Sabena was providing its helicopter services with consistency, and as anyone knows, that is one of the main ingredients of success in the airline industry. If you have no regularity, you have no customers. However, Sabena, which recorded quite exceptional results in its operations during its first year of services, only provided VFR flights. If the visibility was terrible, it had to cancel its flights and transport its passengers by taxi, which was not a cheap option. Thus, the weather hindered Sabena, which relied on stable weather conditions for its helicopter services. That was another chink in the armor of the helicopter services, apart from the costs of providing the service. The main routes which could not afford the cancelation of helicopter flights were the Brussels-Paris and Brussels-London routes, as those two lines were essential to Sabena. The carriers following Sabena's helicopter operations with interest started to think of the viability of helicopter operations. It was no secret that with its four Sikorsky S-55s, whose fares were discounted heavily to attract customers, Sabena recorded heavy losses on its network in the first year. However, because of its geographical location, Brussels was at the center of northwest Europe, in an area where within a 350-kilometer circumference, there were 70 million inhabitants, many of them potential Sabena customers. Sabena was in a unique position and one that it aimed to take full advantage of with its helicopter services. Sabena believed its first-year losses were just a drawback and that its helicopter investment would pay off handsomely. However, as long as Sabena had only S-55s, it could not hope to increase its helicopter operations' profits significantly. Starting a helicopter service was an excellent idea in the beginning. Still, in practice, helicopter operations were not precisely a viable proposition for a national air carrier, especially from one that was not in the habit of making solid profits year in, year out.

In late February, Sabena announced that it had ordered 6 Douglas DC-7C 'Seven Seas'. The DC-7C was the longer-range model of the DC-7 with nonstop transatlantic capabilities, upgraded Wright R-3350-988TC18EA1-2 18-cylinder turbo-compound air-cooled radial piston engines, and additional fuel capacity, primarily in its long wings. The first of the six propliners was expected to be delivered to the carrier in the last quarter of 1956. The DC-7C had a cruising speed of 346 mph (557 kph) and a range of 5,635 miles (9,069 km). Sabena chose it as the airliner to fly the long-distance flights from Brussels to New York and Brussels to the Belgian Congo. Gilbert Perrier, Sabena's Chairman of the board, signed the agreement with Douglas. The Douglas DC-7Cs, with their deluxe interior, offered accommodation for 62 passengers.

In 1955, Sabena declared that its 1954 performance was better than expected. The carrier saw an increase in its traffic numbers which amounted to 84,081,515 ton-km. compared with 73,327,394 in

1953. The number of miles traveled by the company's aircraft increased from 15,691,994 miles (25,253,817 km) to 17,454,537 miles (28,090,355 km).The passengers carried reached 391,756 in 1954, compared to 375,185 in the previous year. That was an increase of 16,571 passengers or 4.4%. Expressed in passenger-miles, it represented a total of 352,459,483, (567,228,555 passenger-km), against 319,437,335 passenger-miles (514,084,560 passenger-km), which was an increase of 10.3%. Sabena's cargo traffic also showed a very pronounced upward curve. In 1954, 23,013,428 ton-km of cargo was transported, compared to 18,579,504 in 1953, which was 4,433,924 ton-km more than 1953, a 23.8% increase. As for mail. Sabena also increased the ton-km transported from 3,738,992 to 4,258,587, 13.8% higher.

Down in the Belgian Congo, on its domestic network, Sabena transported 78,965 passengers. The total was a remarkable achievement, as it was getting closer to the size of the total European population of the Belgian Congo in 1954, which amounted to 80,000. Overall, Sabena's fleet in the Belgian Congo in 1954 consisted of the following aircraft: 13 DC-3, 8 C-47, and 4 DC-4 all-cargo types. The number of Sabena staff in the Congo was 442 European employees and 1,100 Congolese natives. Sabena carried out maintenance of the fleet on site. Leopoldville had well-equipped workshops capable of carrying out all maintenance requirements of the DC-3s and some for the DC-4s. The maintenance workshops of Stanleyville and Elizabethville were smaller and less equipped.

Sabena completed more than 5,000 flights in 1954, including 3,750 on regular scheduled routes, and recorded 7,035 flight hours, including 5,330 on international routes. The total miles flown reached 442,571 miles (712,250 km). On its route network, Sabena opened a new Brussels-Congo line on April 17, with the addition of Paris. The route was Paris-Brussels-Frankfurt-Beirut-Khartoum-Stanleyville-Elisabethville and the east of the Belgian Congo. Incidentally, Beirut became the 87th city serviced by Sabena.

Sabena inaugurated two new routes in 1955. First, on May 18, Sabena opened a Brussels-Rotterdam helicopter service with three weekly flights for the summer period. Then, on July 14, Belgium and Ireland finalized an air traffic agreement after several weeks of discussions. Under the terms of this agreement, Ireland gave Sabena the right of a stopover at two of its most prominent airports, Dublin and Shannon. For its part, Belgium approved the Irish national carrier's request to open a Dublin-Brussels route. Meanwhile, in May, the West German Government granted Sabena the right of a stopover at Frankfurt for its Brussels-Frankfurt-Beirut-Khartoum-Stanleyville-Elisabethville line.

After the success of the 5 Convair 240s in Sabena's service, the Belgian carrier was interested in the newer and sleeker Convair CV-440 'Metropolitan'. On September 2, it placed an order for 12 of the type for USD 17 million. After the Scandinavian carrier SAS, Sabena was the second airline in Europe to order the CV-440s, which featured built-in passenger stairs, improved soundproofing, and provision for nose radar. The delivery of the aircraft was penciled in for the spring of 1956. Sabena had no plans to inaugurate new European routes with the Convair CV-440 'Metropolitans' once they were delivered. It intended to use them to increase the frequencies of its services to the French Riviera, Italy, Switzerland, Austria, and Spain. In the same month, Sabena announced that it had also ordered three more Douglas DC-7Cs, bringing the total to 9 aircraft.

In May, Sabena released its air traffic performance results, which were entirely satisfactory. However, the news was not so good in the financial report released in October. Sabena hit a financial wall in 1954, as it made a massive loss of BFR 53.48 million. It was almost BFR 30 million more than the loss of BFR 24 million in 1953, after the profit of BFR 33.6 million in 1952. Surprisingly, revenue in 1954 increased by 14%. Sabena's President, Mr. Perrier, stated that more than half the amount lost in 1954 was attributable to longer flying hours due to detours of flights that had to fly around areas restricted by military authorities.

Sabena ordered 8 Sikorsky S-58 helicopters capable of carrying 12 passengers. When delivered in the spring of 1956, it would replace its fleet of S-55s, which were smaller than the S-58s and could only carry seven passengers. The S-58 was equipped with a Wright R-1820-84 radial engine with 1,525 hp and was 25% faster than the S-55, with a cruising speed of 112 mph (180 kph) and a maximum speed of 134 mph (215 kph). By mid-November 1955, Sabena had transported more than 50,282 passengers on its helicopter network since the beginning of operations.

The CV-440, DC-7C, and S-58 orders were quite interesting. However, the event which took the imagination of Sabena's staff, and the Belgian people, was on December 28, when the airline announced an order for 3 Boeing 707-329 Intercontinental jetliners. The carrier would use them on its two most important networks, the Brussels-New York and Brussels-Belgian Congo routes. Sabena increased its order to 4 707-329s in March 1956. Flights were due to commence in the spring of 1960. The Boeing 707-329 Intercontinental first flew on January 11, 1959, and Pan American Airways was the first operator of the type. On August 26, 1959, the American giant initially began operating the 707-300 Intercontinental in the United States, and on October 10, 1959, it commenced services across the North Atlantic to Europe. Sabena's 707-329s were scheduled to arrive in December 1959. The airline allowed over one month of familiarization with the new jetliners for the crews before services could commence in January. Sabena's 707s offered accommodation for 126 passengers, 28 in first class, 98 in economy class, and a crew of 9. The Belgian carrier's 707-329s were powered by 4 Pratt & Whitney JT4A-3 turbojets with 16,800-lbs thrust. The aircraft had a maximum speed of 570 mph (917 kph) and a cruising speed of 547 mph (880 kph). Sabena's Intercontinentals would be able to fly from Brussels to New York in just 7 hours and 20 minutes, while a flight from Brussels to Leopoldville would take almost a similar amount of time, 7 hours and 50 minutes. Sabena announced that its 707 Intercontinentals had a payload of 17,375 kg (38,306 lbs.), allowing for 126 passengers' baggage and an extra 5,783 kg (12,750 lbs.) for freight and mail. The 707s had a range of 3,850 miles (6,196 km). The Jet Era had finally arrived for Sabena.

Photo 21. Sabena CV-440 'OO-SCO' at Brussels in May 1966. Delivered to Sabena on October 5, 1956, and sold to Frontier Airlines on November 10, 1967. (Photo by John F. Ciesla)

Photo 22. Passengers are boarding Sabena Convair CV-440-11 'Metropolitan,' 'OO-SCO' at Zurich Airport on May 15, 1966. (Photo by ETH Zurich)

Photo 23. Sabena CV-440 'OO-SCN' at Brussels airport, near the avgas terminal in May 1966. The CV-440 was delivered to Sabena on September 19, 1956, and sold to Frontier Airlines on February 26, 1968. Notice the Belgian Air Force Flying Boxcars in the background. (Photo by John F. Ciesla)

Photo 24. Sabena CV-440 'OO-SCS' at Salzburg Airport in Austria. Delivered to Sabena on November 14, 1956, and sold to Frontier Airlines on March 4, 1968, with reg. 'N73166'. (Photo by Fritz Heusch)

Photo 25. Sabena Convair CV-440 'OO-SCO' at Ostend-Bruges International Airport on September 14, 1963. (Photo by Ken Fielding via Wikimedia)

1956

An extraordinary general meeting of the shareholders of Sabena, which took place on December 13, 1955, dealt with the legally authorized capital increase from BFR 200 million to BFR 500 million for the airline. The Belgian Parliament approved the capital increase by a special law on August 18, 1955, and authorized the government to subscribe to its share. The new shares, valued at BFR 200 million, were immediately snapped up, half by a private group and the other half by the Belgian state and the Belgian Congo. The relationship between private and public participation in Sabena was retained, and the airline continued to operate as a mixed-economy company. Since Sabena was a stock corporation, 50.7% of the voting shares were in the hands of the state. Of the 14 members of the Board of Directors, seven were appointed by the state authorities and seven by private shareholders. The articles of association stipulated that the president, the vice-president, and the member of the board of directors entrusted with the airline's management would be named among the seven members who represented the private shareholders. The amendment to the articles of association was adopted, enabling Sabena to borrow funds under a state guarantee. In 1956 Sabena had an annual turnover of approximately BFR 2 billion, of which BFR 650 million was brought in in foreign currencies.

In February 1956, Sabena released its air traffic performance results which showed a 14.2% increase in air traffic for 1955 compared to 1954. Sabena achieved a total transport performance of 94,897,874 ton-km compared to 83,081,515 ton-km in 1954. The airline offered 133,155,331 ton-km in 1955, almost 11 million more ton-km than the figure in 1954 of 121,849,431. The airline's ton-km increased by 9.3%, while sold ton-km rose from 68.2% in 1954 to 71.3% in 1955. Sabena's aircraft completed 95,966 flight hours and covered 18,385,895 miles (29,589,230 km) in 1955, which was a rise of 5.3%, in comparison to 1954, when Sabena flew 17,454,524 miles (28,090,335 km). The number of passengers carried rose from 391,756 to 449,698, increasing by 57,942, or 14.8%. In its cargo division, Sabena performed very well. In 1955 the company stated that 25,037,216 ton-km of cargo was carried, compared to 22,410,498 ton-km in 1954, an 11.7% increase. As for mail, the numbers showed that in 1955 Sabena carried 5,235,232 ton-km of mail, compared to 4,258,587 in 1954, which was 22.9% higher. In the Belgian Congo, Sabena transported 93,956 passengers on its intra-African network alone in 1955. That was the first time that the Belgian carrier carried 5,000 more passengers on its intra-African route system than the total white resident population in the Belgian Congo. In 1955, Sabena's route network had expanded to 75,633 miles (121,720 km).

In April, Sabena and Aeroflot, the Soviet state-owned airline, signed a unique agreement on the mutual recognition of their tickets. Under the agreement, Sabena was authorized by the Soviet authorities to issue tickets for destinations in the Soviet Union in the future; however, the Belgian carrier was not permitted to fly over Soviet territory. Its passengers, which were heading to the USSR, had to deplane at Prague, then transfer to Soviet planes which would fly them to their destinations, obviously over Soviet territory

On May 6, 1956, Sabena inaugurated a new regular helicopter service, the Brussels-Dortmund route. With this new service, the number of towns in Sabena's helicopter network increased to 12. In the autumn of 1956, the twelve-seat Sikorsky S-58 helicopters gradually replaced the Sikorsky S-55s with which Sabena had been operating regular international passenger services since 1953. Then, in the September winter timetable, Sabena introduced three new European destinations, Warsaw, Vienna, and Budapest. The airliner of choice for all medium-haul routes was now the Convair CV-440 'Metropolitan'. The few remaining cities not serviced by Sabena's CV-440s continued to be served with Douglas DC-6s. A few months later, in January 1957, when the Douglas DC-7Cs arrived on the scene, Sabena used them on its two long-haul routes, Brussels-New York and Brussels-

Leopoldville-Johannesburg. The latest sleek Douglas propliners enabled Sabena to reduce the flight time between Brussels and New York from 16 hours to 13 hours 10 minutes and between Belgium and South Africa from 23 hours 20 minutes to 19 hours 15 minutes. In addition, the more extended flight range of the DC-7Cs would make it possible to have fewer stopovers on the transatlantic and Belgium-Congo routes.

Taking a closer look at the individual countries in Sabena's European route network in 1956, starting with France, the company's Convair CV-440s operated four daily services between Brussels and Paris. There were also five cargo flights a week that the DC-3s handled. In addition, there were five weekly flights scheduled on the Brussels-Nice Riviera route in southern France until December 15. After that, the frequencies increased to 7 per week to meet the demands of the busy Christmas and New Year seasons. There were ten flights a week on offer for Switzerland between Brussels and Zurich. Swissair operated three of the flights, and Sabena seven. Followed by eight weekly trips between Brussels and Geneva, with Sabena taking on the bulk of the flights with five to Swissair's three.

Furthermore, Geneva was now a weekly stopover for the Belgium-Congo route, which increased to twice per week in January 1957. Over in the UK, BEA (British European Airways) and Sabena agreed to operate 33 flights a week between Brussels and London. The Belgian carrier serviced 13 flights until December 1, when they increased to 19 per week. The flight from Brussels to London was nonstop, but there was a stopover at Antwerp on the flight back to Brussels on weekdays. The agreement between the two European carriers also allowed for six cargo flights to be carried out each week between the two capitals. In the Midlands, the city of Manchester continued to be a stopover for Sabena on the Brussels-New York route five times per week. Five flights a week to Rome on the Italian lines were reduced to 4 a week in January due to fewer tourists flying to Italy in the middle of winter. Three of those flights were stopovers on Sabena's Belgium to Congo route. The northern Italian city of Milan was the beneficiary of 6 weekly flights. In Spain, there was only one weekly flight from Brussels to the island of Mallorca during the winter season. In addition, Sabena, in collaboration with the Dutch carrier, KLM, agreed to operate all flights to the Spanish island at the beginning of winter. After that, KLM took over the flights. The Benelux (Belgium-Netherlands-Luxembourg) airlines also conducted a flight to Barcelona. For Portugal, there were two direct weekly flights from Brussels to Lisbon (via Madrid) and two other flights to Lisbon, which were stopovers for the Sabena aircraft flying to the Belgium Congo. On Sabena's Brussels to Amsterdam route, 29 weekly passenger flights were jointly flown with KLM and SAS and five weekly cargo flights.

Concerning Scandinavia, Sabena flew twice a week to Stockholm via Copenhagen. It also served eastern Europe. Czechoslovakia had six weekly connections between Brussels and Prague. The Czechoslovakian national airline, CSA, reserved the right to take over two of the flights in the future. Regarding Poland and Hungary, those two countries became the 34th and 35th countries to be included in Sabena's route network in 1956. Sabena scheduled two weekly flights to Warsaw, and Budapest got two weekly stopover flights on the Brussels-Athens route.

Regarding Greece, Athens was an important city for Sabena. The Belgian carrier's six weekly flights to Frankfurt, Vienna, Beirut, Tel Aviv, Stanleyville, and Elisabethville flew via Athens. Then, from January 1957, Sabena increased the Athens flights to 7 per week. Finally, Sabena offered 20 flights every week from Brussels to destinations in Germany, Austria, and Luxembourg in north-central Europe. Sabena's Convair 'Metropolitans' and DC-6s flew to Luxembourg four times a week, seven times to Dusseldorf, seven times to Frankfurt, four times to Stuttgart, four times to Munich, and four times to Vienna. As for cargo runs, Sabena used its workhorses, the DC-3s, on its cargo services to Dusseldorf and Cologne four times a week and to Stuttgart and Cologne twice a week.

Sabena held its ordinary general meeting of shareholders in October. Gilbert Perier, the Chairman of the Board of Directors, was at the helm. The board approved the balance sheet and the profit and loss account for 1955, which, for a turnover of BFR 2.1 million, showed a gain of BFR 19.3 million. It was not a considerable amount, but it was sufficient. The 1954 financial year showed a deficit of BFR 54 million, which was covered by the Belgian state. The company's income rose from BFR 1.857 billion to BFR 2.14 billion, while expenditure increased from BFR 1.9 billion to BFR 2.1 billion. The number of ton-km offered by Sabena amounted to 133,415,213 in 1955, against 121,849,431 in 1954, an increase of 9.5%. Concerning the number of ton-km sold in 1955, they amounted to 95,181,430 ton-km, which was an increase of 14.6% compared with 83,081,515 in the previous year. The load factor went up by 3.1%, to 71.3%, from 68.2% in 1954.

Meanwhile, Sabena carried 450,000 passengers in 1955, 15% more than in 1954, while passenger-km increased by 14.6%. The company's cargo services rose by 10.1% and accounted for 25.9% of the airline's total tonnage. The report also stated that Sabena made significant but costly efforts to expand its international helicopter network. In the future, the Belgian carrier indicated that it would continue to operate its helicopter services in a more moderate form, even though the total length of the helicopter connections barely made up 3% of Sabena's entire route network. The writing was on the wall; Sabena would scrutinize its helicopter operations in the future. During the 1955 financial year, its capital increased from BFR 300 million to BFR 500 million; however, it did not offer dividends. Instead, the Belgian Government provided substantial financial support for acquiring new aircraft, which required a total expenditure of BFR 1.6 billion. Overall, 12 Convair CV-440 'Metropolitans', 9 Douglas DC-7Cs, and 8 Sikorsky S-58 helicopters were on order by Sabena.

The following month after the general meeting, Sabena announced that in 1955, its expanding air cargo services achieved an upswing of slightly over 10% in ton-km and a revenue gain of more than 13%. As a result of that, on November 1, a new European cargo network was launched by the airline. The new specialized service would be run the same way as Sabena's passenger flights but dedicated solely to air cargo. The Belgian airline's air cargo route network covered destinations in Germany, Austria, Italy, the UK, France, Ireland, the Netherlands, and Denmark. However, Sabena's major cargo lines were the New York-Brussels and Johannesburg-Brussels routes. They were the big money earners for the airline. The Belgian carrier provided cargo connections from its base at Melsbroeck airport in Brussels. Once the cargo arrived in Brussels, it would be carried to many European cities by the airline's Douglas DC-3s. Sabena attached particular importance to its cargo services, which in 1955 represented 25.9% of its total transport volume and 13.3% of its revenue.

Sabena's fleet in mid-1956 comprised the following aircraft: 2 DC-6C, 8 DC-6B, 4 DC-6, 8 DC-4, 29 DC-3, 5 Convair CV-240, and 6 Sikorsky S-55. It had 9 DC-7C, 12 Convair CV-440, 4 Boeing 707-329, and 8 S-58 helicopters on order.

On October 4, 1956, two Sabena Sikorsky S-58 helicopters managed to break a speed record. The two copters took off from Paris bound for Brussels and achieved a speed that, at the time, had not yet been attained by any means of transport. The two helicopters flew from the 'Barnard Heliport' in Paris to the Heliport of the 'Allee Verte' in Brussels, a distance of 193 miles (312 km), in Brussels in just 1 hour, 16 minutes.

Photo 26. A Sabena S-58C at Rotterdam Heliport in 1955. It was leased to Sabena as 'N869' on June 11, 1963, before being bought by Sabena and registered 'OO-SHP' on November 4, 1964. It was sold to the Belgian Air Force on December 13, 1967. (Photo from 'www.douglasdc8.com')

Photo 27. Sabena's S-58 'OO-SHP' on a frigid wintry morning in the Netherlands in the early 1960s. (Photo from 'www.douglasdc8.com')

Photo 28. A Sabena Sikorsky S-58 at Brussels National airport, next to one of Sabena's 707s in 1962. (Photo by Robert Huhardeaux)

1957

Sabena's helicopter operations were expanding. On December 3, 1956, the Belgian company put into service its new Sikorsky S-58s on its Brussels-Antwerp-Rotterdam route. From January 1, 1957, it increased the flight schedule to 12 weekly flights on that route. On January 15, Sabena announced that on March 1, it would expand its helicopter network further with two new lines. The first route was the Brussels-Paris route, which would offer five daily flights. The departure point in Brussels was the 'Alle Verte' helicopter port, while the arrival point in Paris was at the 'Issy-Paris, Place Balard' heliport. The flight would take 1 hour and 45 minutes. The second line was the Brussels-Luxembourg-Saarbrucken route, on which Sabena provided daily flights on the route. With the opening of these two routes and the Brussels-Dortmund route on May 6, 1956, Sabena increased the number of cities included in its helicopter network to eleven.

In February, Sabena announced a new record. For the first time in its history, the number of passengers transported on its route network in 1956 reached 526,594. In contrast to the recent past, when Sabena suffered many losses, the previous year was a good year for the airline as it made a considerable profit of over BFR 123 million. After deduction of BFR 45 million tax provision, the net profit was BFR 78 million. In 1956 Sabena's ton-km figure was 109,495,963, compared 95,181,430, in 1955, an increase of 15%. The number of passengers carried reached 526,594, 17% more than in 1955. The number of ton-km of freight increased from 24,671,418 in 1955 to 27,780,341 in 1956, a rise of 12.6%. Moreover, the airline's load factor reached 74.2% against 71.3% in the previous year. Meanwhile, the aircraft in its fleet covered 19,825,747 miles (31,906,448 km) against 18,436,503, miles (29,670,677km) in 1955, and the total flight hours in 1956 reached 102,471, against 96,250 in the previous year. Mr. Deswart, Sabena's Managing Director, emphasized that in 1956 Sabena expanded its air services substantially.

Early in 1957, Sabena flew to New York nine times a week. Seven of those flights were on the DC-7Cs, and the other two were operated by the DC-6As, on cargo duties. When Sabena released its summer timetable for 1957 in March, the carrier revealed that its North American flight frequencies to New York would go up to 12 weekly flights from June 1. Then they would increase to 14 weekly flights during the two summer peak periods from June 23 to July 14 and from August 11 to September 9. The two transatlantic services included a stopover at Shannon, Ireland. Meanwhile, the number of weekly stopovers in Manchester gradually increased from 5 to 7. Sabena managed its timetable very well that summer, as it offered five weekly flights to New York, with a 12.30 pm departure from Brussels at 12.30 p.m. and a 9.30 pm arrival on the same day (New York time), in the Big Apple. That allowed the passengers to head straight to their hotels to sleep off their weariness from the flight. In Europe, the Brussels-London route was extremely busy with 41 weekly flights, which Sabena and BEA shared. The frequencies intensified to 45 per week between June 15 and September 30. Manchester benefited from Sabena's schedule, as there were seven to eleven weekly flights on offer from Brussels to Manchester each week. Five of them were transatlantic flights, and the two others were operating the new Brussels-Manchester-Dublin route. Furthermore, two other Brussels-Dublin services via Manchester were provided by the Irish national carrier, Aer Lingus.

In France, Nice was the recipient of daily connection to Brussels throughout the summer season. From June 15, Sabena added four extra flights to increase the number of flights to 11. A total of 45 flights were made available during the summer by Sabena on the following routes: Brussels to Salzburg two per week, Munich four per week, Vienna five per week, Luxembourg and Stuttgart four both four per week, Frankfurt seven per week; Hamburg 11 per week, and Düsseldorf 12 flights per week. The frequency of the Brussels-Zurich line, which Sabena and Swissair provided, was 12 per week, while the two airlines flew daily to Geneva. Sabena offered seven flights to Rome each week; two of those flights went via Milan, two continued to Naples, and three stopped over in Rome on the way to the Belgium Congo. The capital of Belgium's northern neighbor, Amsterdam, benefited from 25 weekly flights. Sabena operated those flights with its Dutch and Scandinavian copartners, KLM and SAS. Sabena provided two weekly flights on the Scandinavian network: one on the Brussels-Amsterdam-Copenhagen line and another one on the Brussels-Amsterdam-Copenhagen-Stockholm route. Sabena and KLM operated four weekly flights from Brussels to Barcelona in the Mediterranean, two by Sabena and two by KLM. Four other flights were made available on the Brussels-Palma de Mallorca route, two by Sabena and KLM. Due to its ideal location on the way to the Near East and Congo routes, Sabena's important south European stopover, Athens, received seven weekly flights, which increased to eight from June 17. Sabena opened a weekly Brussels-Vienna-Athens-Tel Aviv service. Then the Belgian airline began a twice-weekly Brussels-Beirut service via Frankfurt and Athens in the summer. Over in Eastern Europe, Sabena scheduled seven weekly flights to Prague, two by the Czechoslovak airline, CSA, and five by the Belgian company. There were nine weekly flights between Belgium and Poland during the summer of 1957, 6 by the Polish airline LOT and three by Sabena. In addition, Sabena inaugurated an Eastern European line on March 13, Brussels-Budapest-Bucharest, with two flights a week. Sabena's Belgian-Congo flights averaged 17 per week in the African network.

On the Brussels-Leopoldville line, it was between nine and ten. The Brussels-Stanleyville route between six and eight, and two on the Brussels-Stanleyville-Elisabethville line. The stopovers for the Congo network were in the following cities: Rome, Athens, Tripoli, Cairo, Khartoum, and Libenge (Congo) on the Eastern route, and Geneva, Rome, Lisbon, Casablanca, and Tripoli on the Leopoldville line. In addition, two flights on the Brussels-Leopoldville route continued their flights to Johannesburg.

Sabena's helicopter network in 1957 saw increases in the frequency of flights between Brussels and Paris, which increased from two to five per day. The Brussels-Paris route was inaugurated on March 4, 1957, and during the first month of operation, 44 helicopters flew the route and carried 692 passengers. Sabena operated the Brussels-Lille line twice daily and the Brussels-Rotterdam line via Antwerp three times a day. Moreover, the carrier scheduled three daily flights on the Brussels-Cologne connection via Liege and Maestricht. Two of those flights continued to Bonn. Eindhoven, Dortmund, and Duisburg also benefited from three daily flights. Sabena's flight operations increased significantly in the summer of 1957 compared to its summer timetable of 1956. The reason was that the Belgian carrier had many more aircraft in 1957.

June and July saw Sabena reach two milestones. In June, the Belgian company carried its 100,000th helicopter passenger on a Sikorsky S-58 flying from Brussels to Bonn. Then in July, arguably the more significant milestone occurred on July 8, when a Sabena DC-6B left Brussels at 7 pm for Elisabethville on the overnight flight. It was the carrier's 5,000th flight to the Congo. The first flight to the Congo took off from Brussel's Haren airport for Leopoldville on February 23, 1935. The famous Belgian aviator Prosper Cocquyt was at the controls of the Fokker airliner, which carried 13 mail bags and a significant passenger, Tony Orta, the Director of Sabena's operations in Africa. The Fokker trimotor landed at Leopoldville after five and a half days. By 1957, Sabena's modern fleet of DC-6s and DC-7Cs completed Belgium to Congo flights in about 16 hours. It had taken Sabena 10 years to double the milestone of the first 1,000 services on the Belgian-Congo route network. However, it only took 17 months to complete the last 1,000 flights. In the 22 years since the first flight to the Congo, Sabena completed 5,000 flights, carried over 351,000 passengers, and covered 43.5 million miles (70 million km).

Sabena's ordinary general meeting of shareholders was held on October 1, 1957, chaired by Gilbert Perier, Chairman of the Board of Directors. In the assembly, he stated in 1956, Sabena recorded a net profit of BFR 77.4. The previous year's net profit was BFR 19.32 million. The shareholders approved the balance sheet and the profit and loss accounts for the 1956 financial year. The company's transport revenue figure of BFR 2.3 billion was 17.5% over the previous financial year, resulting in a net operating profit of BFR 77.4 million. A significant increase in the ton-km offered and sold influenced the excellent results for 1956. The report indicated that the ton-km offered was 147,509,242 in 1956, against 133,415,213 in 1955, which increased by 10.6%. Meanwhile, the number of ton-km transported increased from 95,181,430 in 1955 to 109,495,963 in 1956, a 15% increase. The load factor also rose in 1956 to 74.2%, which was the highest percentage recorded by Sabena since 1946, compared to 71.3% in 1955.

Moreover, in 1956, Sabena's aircraft transported 526,594 passengers, 16.9% more than the previous year. Cargo transport increased by 12.6 %, representing 25.9% of the airline's total capacity. The helicopter network, which was the only one of its kind globally, carried over 17,000 passengers in 1956, a growth rate of 5.3% compared to the previous year. Finally, at the Annual General Meeting, Mr. Gilbert Perier was re-elected as Chairman of the Board of Directors.

On October 10, 1957, Sabena expanded its route network after inaugurating a new twice-weekly Belgrade route. The carrier's helicopter operations were also moving forward. Sabena provided 12 weekly flights from Brussels to Paris. The route had been successful since Sabena inaugurated it on March 1. For example, in August alone, 1,503 passengers were transported on 111 round-trip flights between the two capitals, with an average of seven passengers per flight. On the Brussels-Lille line during winter, Sabena provided six weekly flights with the S-58s. Montreal was penned in as a stopover on the Brussels-New York route twice a week. Istanbul was connected weekly to the Belgian capital via Frankfurt and Athens. In addition, Sabena tried to improve air links between

Belgium and eastern Congo. It provided one flight to the region until March 15, 1958, and then offered a second flight, with a DC-7C connecting Brussels and Elisabethville, via Cairo. The Belgian carrier also provided a once-a-week flight to Cairo via the following route; Brussels-Prague-Athens-Cairo. Interestingly, in 1958, Sabena reduced the frequency of its regular helicopter flights between Brussels and eleven cities in Belgium, France, the Netherlands, and West Germany.

By the end of December 31, 1956, Sabena's fleet consisted of 86 aircraft: 3 Douglas DC-7C, 8 DC-6B, 2 DC-6 A, and 3 DC6, 8 DC-4, 12 Convair CV-440 'Metropolitan', 4 Convair CV-240; 28 DC-3, C-47 and C-49 types; 4 Sikorsky S-58 helicopters, and 14 trainers. In addition, it had 7 DC-7Cs on order. In 1957, Sabena increased its S-58 fleet to 8 helicopters. Meanwhile, its training division increased its Cessna 310 light aircraft fleet to nine aircraft throughout 1957. Sabena's number of employees at the end of 1955 was 8,057 people, which increased to 8,746 by the end of 1956.

On June 4, 1957, Sabena conducted an interesting minor operation collaborating with Air Charter Ltd, a UK charter airline, which later changed its name to 'Channel Air Bridge.' The two airlines agreed to operate a regular passenger and vehicle air service between Southend, on the English Channel coast, and Ostend in Belgium. The services ran at six per day, then 12 in both directions from July. The Bristol 170 Mark 32 twin-engine propeller freighter was the aircraft that flew the route. The Bristol 170 could carry either four small cars, three medium-sized cars, or two large cars and 16 passengers plus bicycles, motorcycles, and freight. The flight time between Southend and Ostend was 40 minutes.

Photo 29. Bristol 170 Superfreighter 32 with UK reg. 'G-APAV' owned by Channel Air Bridge but painted in Sabena colors, taking off from Southend on November 25, 1964. This aircraft was the last Bristol 170 Freighter produced. Sabena only ever flew the Bristol 170 Freighters in collaboration with 'Air Charter Ltd/ Channel Air Bridge.' It never contributed its aircraft to the Channel services.
(Photo by Richard Goring via Wikimedia, colorized)

1958

In 1957, Sabena transported 686,000 passengers. The figure was 30% higher than the total number of passengers in 1957, which was 526,594, and was also the highest growth rate in passenger numbers since 1946. The ongoing modernization of Sabena's fleet was instrumental in increasing passenger capacity, resulting in higher passenger numbers. Sabena's fleet at the beginning of 1958 consisted of 77 aircraft: 6 DC-7C, 8 DC-6B, 3 DC-6, 2 DC-6A, 12 Convair CV-440, 4 Convair CV-240, 9 DC-4, 28 DC-3, and 8 S-58 helicopters. There were still 4 DC-7C to be delivered, and, more interestingly, its new jetliners, the 707-329s, were due to arrive in 1959.

Sabena's number of ton-km offered rose from 109,495,963 ton-km in 1956 to 137,747,670 ton-km in 1957, an increase of 29%. The ton-km used figure was up 26%. The remarkable increase was primarily due to the commissioning of the Douglas DC-7Cs, which operated Sabena's long-haul networks. Sabena made great progress on the transatlantic route to North America. The carrier increased its passenger numbers by 60% on that route network, and its planes flew over the Atlantic 1,036 times, compared to 787 in the previous year. On the other hand, the Belgium-Congo line expanded slower than in previous years. The total number of passengers in 1957 was 73,143, which was only an 8% rise compared to 1956.

Sabena was taken aback in 1957 when its Belgium-Congo route developed at a slower rate than in prior years. As a result, the number of passengers transported on that network in 1957 was 73,143, increasing just 8%. In previous years, the gains were in the double digits. The good news for 1957 was that the Belgian company's DC-7Cs, with their increased capacity, began flying the two long-haul routes, Brussels-New York and Brussels-Leopoldville-Johannesburg, at the start of the year. As a result, Sabena transported 128,218 passengers to the Belgian Congo in 1957, or 20,000 more than the whole population of Belgium's overseas territories.

In 1957, Sabena's ton-km offered skyrocketed by 48%, compared to 1956, while the used ton-km increased by 37%. In Europe, the number of passengers flying with Sabena rose from 302,700 in 1956 to 395,400 in 1957. Meanwhile, Sabena increased its European helicopter network's ton-km to 188, while the number of passengers using its services rose from 18,474 in 1956 to 40,871 in 1957. The trendy route in 1957 was the Brussel-Paris line, which Sabena opened in March 1957. In less than a year, Sabena transported 11,541 passengers on that service. In Europe and the Near East, Sabena expanded its route network considerably. Sabena added the following cities to its schedules in 1957; Europe: Belgrade, Bucharest, Budapest, Dublin, Istanbul, and Southend; North America: Montreal, Africa: Gemena (Belgian Congo), and Salisbury (in Rhodesia, now known as Zimbabwe).

The summer timetable was released in February, which showed a significant increase in Sabena flights to and from Brussels due to the 1958 Brussels World Fair held in the Belgian capital. The timetable was effective from April 1 until October 31, 1958. The Brussels World Fair placed heavy demands on Belgian air traffic, as Sabena reinforced almost all its routes originating from Brussels. Fortunately, Sabena increased its fleet over the previous two years due to purchasing 12 Convair CV-440 'Metropolitans' and 10 Douglas DC-7s, which made things easier. The carrier increased the frequency on the North Atlantic route to 3 daily flights, and, on certain days, it added extra flights. From the Belgian Congo, there were 19 weekly flights to Brussels. The Brussels-London route was provided with 11 flights a day, whereas, on the Brussels-Paris line, there were six daily flights. In addition to the fixed-wing flights, Sabena's helicopter service between the two capitals provided five daily flights in both directions. Sabena's international helicopter network provided flights from other European cities directly to the world fair. A helicopter landing pad was set up near the exhibition's amusement park. The helicopter flights represented the fastest way to reach the fair for the exhibition's visitors. Even though the Brussels World Fair was an important event for Belgium, Sabena also had to think about its network.

An agreement was signed on January 21 between the states of Belgium and Morocco. The contract allowed Sabena to open a direct route between Brussels and Rabat and to continue that service to Leopoldville. The national carrier of Morocco, 'Royal Air Maroc' (RAM), flew the direct Rabat-Brussels line. The first flight between the two nations was opened in the Summer. There were plans to also open a Brussels-Madrid-Lisbon-Casablanca route in the future. From June 2, Sabena was set to open a weekly nonstop Brussels-Moscow flight with its Douglas DC-7C aircraft. This time around, however, Sabena could fly over Soviet territory. The weekly flights were to end on July 31, as from August 1, Sabena increased the frequency to twice weekly. The flight time between Brussels and Moscow was 5 hours and 35 minutes. The Russian national airline, Aeroflot, flew to Brussels once a week, on Fridays, with Tupolev Tu-104 twin-engine jetliners, which completed the journey in just 3 hours. However, the DC-7C was no match for the Red Rocket from Moscow. On November 1, Sabena added another two cities to its network, Ankara, and Tehran. The Ankara flight was once a week from Brussels via Frankfurt, Athens, and Istanbul, while the second to Tehran was once a week. It was the same route as the Ankara one, except Tehran was its final destination.

On April 9, a law was passed by the Belgian Government authorizing Sabena to increase its capital from BFR 500 million to BFR 700 million before the end of 1959 and to BFR 1 billion after January 1, 1960. In addition, the Belgian state guaranteed the granting of loans to Sabena of up to BFR 4 billion, which was the prescribed limit. The draft law was related to financing new airliners for the national carrier. The double increase in capital was subject to a decision of the general meeting of shareholders of the Belgian Company. By the statutes of the latter, half of the subscriptions had to be made by the private sector and the other half by the Belgian State, the Belgian Congo, and 'Rwanda-Urundi,' a colonial territory bordering the Belgian Congo, which Belgium ruled from 1916 to 1962.

On October 1, the ordinary general meeting of Sabena shareholders was held under Mr. Gilbert Perier. The shareholders approved the balance sheet and the income statement for the 1957 financial year. Sabena's revenue rose in that year by 23.6%, with a total of BFR 2.85 billion, and the company achieved a net operating profit of BFR 55.7 million. A significant rise in the ton-km offered and ton-km sold in 1957 affected the results favorably. For the first time, the number of passenger kilometers topped one billion. The number of ton-km offered increased from 147,509,242 in 1956 to 191,059,208 in 1957, a 29.5% rise, while the ton-km transported increased from 109,495,963 in the previous year to 137,747,670 in 1957, a 25.8% upswing. Sabena made significant gains on the North Atlantic and the Middle East networks. The Belgian airline improved its performance in these two networks by 53.8% and 46.3%, respectively, compared to 1956. The helicopter network performed even better than the previous year, increasing by an incredible 205.9% ton-km. Sabena's airliners transported 686,090 passengers in 1957, representing a 30.3% increase over 1956.

Furthermore, the number of passenger kilometers grew by 39.3%. Freight transport increased its share by 11% to 22.1% of the total tonnage handled by the firm. In conclusion, Sabena's helicopter network carried 39,661 passengers in 1957, compared to 17,157 in 1956.

Sabena's 1958 winter timetable had a few changes. First, the company decided to keep the same frequencies as those recorded during the summer of 1958 on the Belgium-Congo route. Then, from March 10, 1959, Sabena offered six weekly services to Stanleyville and three weekly flights between Brussels and Elisabethville. Furthermore, Sabena extended one of those flights to Johannesburg via Salisbury (Rhodesia). In addition, the Belgian carrier extended one of the eight weekly services from Leopoldville to Luanda (Angola). In Europe, flights operating to Palma de Mallorca would stop over in Barcelona. Elsewhere, Vienna became a stopover on the Brussels-Athens-Cairo route.

Photo 30. A group of Air Hostesses walks through the Brussels World Fair in 1958, with the famous Atomium structure in the background. (Photo by Pieter.vandeweyer via Wikimedia)

Photo 31. This Lockheed L-1049H was leased by Sabena from Seaboard & Western Airlines from May until October for the 1958 World's Fair in Brussels. Registration 'N1007C'. (Photo by John F, Ciesla)

1959

The year of the Brussels World Fair in 1958 helped boost Sabena's image significantly. The carrier registered almost 1 million passengers in 1958, an increase of 300,000 compared to 1957, which was previously the most robust year in passenger numbers, reporting 160,000 more passengers than in 1956. Sabena's participation in the exhibition gained the carrier many new friends. The Brussels World Fair's committee built an ultra-modern heliport in the Brussels World Fair's surroundings. That allowed up to 15,000 passengers from Germany, France, and Holland to use Sabena's helicopter services to fly to and land inside the exhibition area. A further 65,000 passengers took the helicopters to fly over the exhibition and take in the scenery from above. The 80,000 people who used Sabena's services would have had fond memories of flying on Sabena's whirlybirds. Sabena was on cloud nine in 1958, as it also opened three new routes in its network, Moscow, Ankara, and Nigeria.

Sabena's summer timetable, which came into force on April 1, 1959, confirmed that Sabena would not slow down after the excellent results in 1958. The carrier included Glasgow, the industrial and commercial center of Scotland, as the latest city in its service. The airline's route network expanded to 124,274 miles (200,000 km) in 1958.

On June 5, 1959, Sabena established a weekly flight between Glasgow and Brussels, and on June 21, it increased the flights to two a week. All in all, Sabena provided 621 flights for the summer season. Moreover, the number of transatlantic flights increased slightly compared to the previous summer. To meet the increasing demand for seats that season, 24 weekly flights, three per day, were made available on the North Atlantic route from June 15. In Europe, Sabena notably increased the frequency of flights to the following destinations: Luxembourg, Nice, Milan, Rome, and Dusseldorf.

Sabena's shareholders' ordinary general meeting was held on July 22, 1959. In 1958, Sabena made a turnover of BFR 3.58 billion, an increase of 16.9% over the previous year. However, the operating deficit was BFR 120.3 million, or 3.4% of the company's turnover. The Board of Directors' report stated that the company's results were adversely influenced by a reduction in the load factor, which dropped from 72.1% to 69%. In addition, the increase in operating expenses, which increased in proportion to production and revenue, was mainly due to the burdensome conditions of operating air charters in 1958 to meet the temporary increase in traffic caused by the 1958 Brussels World Fair. However, there was a silver lining; those air charters made it possible for Sabena to recruit temporary staff, thus avoiding recruiting permanent personnel and buying extra equipment. In 1958, Sabena recorded a significant increase in its traffic. The ton-km sold saw a 16.7% rise, from 137,747,670 in 1957 to 160,801,267 in 1958. In addition, there was a 39.1% increase in the number of passengers carried, 954,630 against 686,090 in 1957. That was the largest percentage increase in 10 years. Cargo transport continued on an upward trend, although in 1958 it was less pronounced than in previous years. Sabena registered a slight increase in cargo traffic of 2.7% over 1957. Cargo carriage accounted for more than 19% of the company's total tonnage transported. The report also affirmed that in 1958 Sabena carried 50,715 passengers on its regular international helicopter network, which was 27.9% more passengers than in 1957.

December 1959 was a busy one for the Belgian carrier. On December 14, Sabena placed an order to acquire 4 Caravelle VI jetliners from the French aircraft manufacturer Sud-Aviation. Sabena was to be the launch customer for the Caravelle VI model, and the first delivery to the airline was to be made in January 1960. Sabena also obtained an option for four aircraft of the same type. The Caravelle VI differed from the previous versions of the Caravelle in that it had a more extended range and a higher payload. The Caravelle Series VI-N was powered by two Avon RA.29 Mk. 531 turbojets with 5,535 kg of thrust each and had a cruising speed of 511 mph (822 kph) at 30,000 feet. The French jetliner's maximum range was 1,410 miles (2,270 km). However, Sabena's model would have fuel tanks in the wings with a capacity of 18,500 liters, which would increase the aircraft's range to 1,553 miles (2,500 km). In addition, Sabena planned to place the Caravelles on its European and Middle Eastern route networks. On Monday, December 21, 1959, exactly one week after placing the order of the 4 Caravelle VI jetliners, the company's first jet airliner, the Boeing 707-329 Intercontinental, which was flown by a Belgian crew, landed in Brussels at 10.15 am. Sabena officially joined the 'Jet Age' on that glorious day.

AIR UNION

In February 1959, the airlines of five of the inner six countries in the European Economic Community (EEC), formed two years earlier, met in merger negotiations. The airlines involved in the talks were Sabena, KLM, Air France, Lufthansa, and Alitalia. The only airline that was not in the negotiations was the airline from Luxembourg, Luxair. Their talks were focused on jointly managing the aviation services of the EEC.

Then on March 1, the representatives of the national airlines of the six countries of the common market met in Paris for discussions. They all agreed they needed to move towards closer cooperation. The reason for wanting the amalgamation of their airlines under the name of 'Air Union' was to be seen as a powerful combination in Europe and help all airlines involved achieve greater profitability. The Air Union would also create a means of power with which the EEC carriers could face their American rivals more effectively in competition. The ultimate goal was to form an umbrella company, similar to the Scandinavian Airlines System's one (SAS), in which the three Scandinavian countries conjointly operated SAS. In the meeting, a proposal was made that the five European airlines combine their resources and fly the transatlantic route as a union with the group's jetliners, just as SAS was doing.

In late March 1959, a Sabena official stated that the negotiations were going as scheduled. In the talks, it was proposed that the new airline would be called 'Europair.' In Europe, in 1951, 44 European and east European airlines from 18 different countries flew 340 flights. However, they shared a combined air traffic volume the size of which could not even match the total volume attained by American Airlines alone in 1950. Thus, Europe was up against titans, like United, Pan Am, and TWA, and to beat any titan, you have to fight as one, but would Air Union be the answer? Moreover, European countries were not exactly renowned for collaborating as one, with Scandinavians being the exception.

Differences of opinion surfaced in early April, with the issue being the quotas of the individual companies in the Europair pool. The difficulties in solving this problem lay in the different levels of development of the five airlines. In the 1950s, the Air Union's two minor airlines, Lufthansa and Alitalia, were finding their feet. As a result, they both supported the merger. On the other hand, KLM was the second-largest airline in the Air Union, behind Air France, and had an expansive route system. Air France and Sabena were regarded as two leading airlines with years of experience in Europe and overseas. In a merger, both airlines would divert part of their capacity to service the routes to their overseas territories. The Europair pool saw its first departure from the union in April. KLM withdrew from the Air Union after a disagreement with the other four airlines.

KLM wanted to organize the collaboration with its partners before the establishment of the quotas, which it stated could be fixed later. All four airlines replied in unison, quotas first, then final organization. They said that KLM did not commit itself to a quota distribution. KLM, having been in existence for 40 years and was the oldest airline in the world, felt strong enough to reject the Europair pool traffic quotas, as it determined they were inadequate. The KLM delegation packed their bags and left. The Euro pair pool would extend to the routes flown by the airlines mentioned in Europe and to countries outside Europe, but not to the colonial areas, which was a good thing for Air France and Sabena. However, the Europair pool could not afford to provide non-member airlines more access to Europe than they had already granted to foreign airlines in individual nations. Large American airlines, in particular, would face restrictions on their rights on the European continent, which they would be unlikely to accept without protest.

On May 20, 1959, the four airlines in the Europair pool, Sabena, Air France, Lufthansa, and Alitalia, reached an agreement in Paris on close cooperation in the long-haul flights, especially on the trans-Atlantic route. They agreed on provisional transport quotas for each of the four national airlines and the sharing of costs. They also discussed which flight routes Europair should fly. Several sub-committees were entrusted with working out the details of the new Europair pool. After the Paris meeting, it was revealed that the quotas for all four airlines were as follows, Air France 31% (the other French airline UAT-TAI got 3%), Lufthansa 30%, Alitalia 26%, and Sabena got 10%. The new kid on the block, the 4-year-old Lufthansa, came off very favorably with a 30% quota. Lufthansa got the high allocation due to a joint venture it agreed with Air France. Without the French carrier's support, Lufthansa would have been lucky to have reached a 10% quota. Many in the aviation world surmised that the resistance of KLM, which left the Europair pool, was mainly directed against the Lufthansa quota, which the Dutch delegation believed it did not deserve.

On May 27, 1959, the Presidents of Air France, Alitalia, Lufthansa, and Sabena, simultaneously announced the agreement reached between their airlines would achieve complete commercial solidarity in the operation of their international networks. The fundamental principles of the deal were: 1) The identities and the structure of the four companies would each retain their staff and resources that they would make accessible to Air Union. Similarly, all agreements they entered with companies that were not part of Air Union remained valid. Furthermore, the airlines retained their freedom to conclude contracts with other airlines. 2) On international lines, there was complete commercial unity. After subtracting its expenditures, Air Union would divide the revenue among the airlines that delivered the aircraft in flight condition by the seats-km or ton-km offered. 3) Open door policy. The new agreement was widely open to any interested airline, European or not 4) The language used within Air Union would be the French language. The problematic question of the distribution of production and the distribution of revenues was resolved as follows: The four airlines' activities in 1958, which, combined, totaled 489 million ton-km, served as the starting point. Increases in all the carriers' output in the future would consider the successes they accomplished in recent years and what they may realistically expect in the future. By the time the aggregate of their 1958 traffic had expanded fivefold to 2.4 billion ton-km, these considerations had allowed Air Union to calculate each carrier's portion.

After more than two and a half years of gestation, negotiations for the creation of the Air Union reached a stage of gridlock. Several facts made many officials seriously doubt the participants' practical solidarity of interests. First, there was the conclusion of a long-running discussion between the Federal Republic of Germany and the Scandinavian nations about SAS traffic rights in Germany. Lufthansa's Air Union partners requested that Lufthansa guarantee the German market would be adequately safeguarded against SAS and KLM's positions. The goal was to achieve some parity across the various markets served by Air Union member airlines. The Germans offered assurances. When the matter reached the level of high politics, it became clear that Germany's significant business interests in Scandinavia exceeded the desire to maintain German air transport's position against the Scandinavian carrier. In the circumstances like this, it was challenging to advocate an actual aviation strategy. Air Union was becoming less of a dream for some.

The years passed, and no firm decisions were made on Air Union. Finally, in 1965, it was all but over. The group of four airlines had attempted, for five years, to create an airline that would have placed it in the top ranks of the world's airlines. The lengthy discussions that Air France, Alitalia, Lufthansa, and Sabena had would have led to the development of a commercially profitable project that could be operated judiciously after gradual adaptation. However, the French government's opposition to endorsing the plan was the leading cause of Air Union's failure. Finally, in 1964, the Foreign Ministers of the four countries decided to re-examine the possibility of forming the Air Union, provided that it became part of the European Common Market, which postulated, in particular, the inclusion of KLM and Luxair. However, by 1965 the progress made was slow at best. After all the promises, meetings, negotiations, and dreams, nothing came of it in the end. Air Union was dead in the water.

Photo 32. Sabena 707-329C 'OO-SJO' at Salzburg W.A. Mozart Austria, on January 26, 1978. Delivered to Sabena on December 3, 1969. Sold to Scibe Airlift, from the Democratic Republic of the Congo, on January 30, 1985. (Photo by Michael Bernhard)

Photo 33. This 707-329C, 'OO-SJL,' was delivered to Sabena on September 30, 1968. It eventually ended up with NATO in the mid-1980s. Here, it is on finals at London Heathrow in July 1977. (Photo by Eduard Marmet)

Photo 34. Sabena 707-329 'OO-SJG' is seen here at Frankfurt am Main airport in June 1976. Delivered to Sabena on February 30, 1960, and was sold to Sabena's subsidiary Sobelair in May 1974. (Photo by Eduard Marmet)

Photo 35. Sabena 707-329C 'OO-SJM' on finals at London Heathrow in April 1974. It was delivered to Sabena on June 17, 1969, and remained with the carrier until November 26, 1979, when it was also sold to Sabena's subsidiary, Sobelair. (Photo by Richard Vandervord)

Photo 36. This is another photo of 'OO-SJM' in Sabena's classic livery. It is seen here at Tokyo Haneda International airport in August 1972. (Photo by Nobuo Oyama)

Photo 37. 707-329 'OO-SJA' was delivered to Sabena on December 4, 1959. This 707 was the first received by the airline. It was eventually sold to Sobelair in January 1978. It is seen here at Tenerife Norte, Los Rodeos (TFN / GCXO) in 1975. (Photo by Bernhard Bauske)

Photo 38. 707-329 'OO-SJM' is about to land at a cold and snowy Munich airport in January 1973. (Photo by Wilhelm Hell)

Sabena Expands
1960-1969

1960

The beginning of the new decade was not a joyous time for Belgium. The Belgian Congo was in flames, with the indigenous population fighting for independence from the white European settlers. The events in 1960 in the Belgian Congo will be examined at the end of this year. But, first, we will delve into Sabena's performance and other details in 1960.

Sabena finally had its jetliner, the beautifully streamlined 707-300 Intercontinental model, delivered to the carrier on December 4, 1959. Just a month later, on January 23, 1960, Sabena inaugurated the first Brussels-New York service with its brand new 707. Sabena created a record with that flight, as it was the first airline in the world to provide a 707-329 jet service to the USA by a non-US-based airline. Sabena scheduled two weekly flights to New York with its Intercontinental jets, which offered accommodation for 130 passengers, 96 in economy and 34 in first class. However, that was not the first service flight by Sabena's 707-329. On January 19, the Belgian carrier's first actual regular service with the new jetliner began when the 707 'Intercontinental' flew the Belgium-Belgian Congo (Leopoldville) route. With the 707-329, the other new service was to Johannesburg, inaugurated two days after the New York service, January 25, 1960. The frequencies of the three new long-haul services, operated by the 707 until March 31, 1960, were once weekly to the Belgian Congo and South Africa and twice weekly on Sabena's North American network.

To make things a bit interesting for the two cold war superpowers, on April 6, 1960, a Sabena Boeing 707-329 'Intercontinental' established the first jetliner link between New York and Moscow. The Belgian carrier's new weekly direct flight from New York to Moscow was with a stopover at Brussels. Sabena operated the same aircraft between the two cities without switching aircraft in Brussels. The first flight was another record for Sabena and its 707, as it was the first time a US-built jetliner served the Russian capital on a scheduled service. The total flight time for the route, including the stopover at Brussels, took 9 hours 45 mins. The aircraft flew back to New York via Brussels and Montreal on the return trip from Moscow.

Sabena's performance in 1959 was not a pleasant one. The Belgian airline suffered a loss of BFR 197 million. Nevertheless, the airline's Managing Director, Mr. Deswarte, declared that the company had managed to bring the 1959 deficit below the ceiling for State intervention, which was BFR 260 million. The unsatisfactory financial results of 1959 to 1960 were attributed to three events: the incidents in the Belgian Congo, which began in January 1959, and the subsequent halt in the Congo's economic expansion, which inadvertently caused the stagnation of air traffic on the Belgium-Belgian Congo route. There was also the setback of the 'Brussels World Fair' in 1958, which created intense competition for Sabena from North Atlantic airlines flying to Brussels for the world fair. The competition undermined Sabena's strength in its territory and kept it from making a financial windfall. Then, there was the slow recovery of the Belgian economy following the worldwide 1958 recession. All three factors created a significant disillusionment for the Belgian carrier in 1959.

No doubt, 1959 was a bad one; however, the company's Board was optimistic. Mr. Deswarte stated that the world economic recovery, which took place from the middle of 1959, created an increase in Sabena's air traffic numbers, which were up by 17% on all networks by the end of 1959. The rapid commissioning of Sabena's Boeing 707-329 'Intercontinental' jetliners, which the airline utilized efficiently, created high passenger traffic volumes, especially on the Johannesburg and North Atlantic routes, where the company was making hay before other airlines introduced their 707 jetliners on Sabena's routes. In May, a bill amending Sabena's statutes was approved by the Belgian Cabinet Council. The explanatory memorandum included a justification of a 65% majority stake by the Belgian State in the capital of Sabena due to the upcoming constitution of the Air Union and the special status of the other three members airlines of that association. The quota allocated to Sabena by the Air Union was 10%, not including traffic between Belgium and the Belgian Congo, Rwanda-Urundi, and Congolese domestic routes. The modification of Sabena's articles of association would transfer an essential part of its activity to Air Union, resulting in significant delegations of power. The state considered that it had to strengthen its abilities within Sabena. Thus it increased its stake in Sabena from 49% to 65%. Whereas that of the Belgian Congo and Rwanda-Urundi would remain unchanged at 25%. Meanwhile, that of the private sector would be reduced to just 10%. Due to the loss of commercial autonomy of Sabena and its dilution within the Air Union, the private sector accepted the takeover of a significant part of the participation in the new company's capital by the State. In addition, Sabena's board of directors would include 15 State members, nine of whom would be appointed by the government, six by the communications department, two by finance, and one by foreign affairs. The private sector would have only two members, while the Belgian Congo and Rwanda-Urundi would have four. Overall, Sabena continued to be considered a mixed company.

In June, Sabena announced that it expected a 40% increase in its scheduled operations in 1960, after its promising performance in 1959 when it carried 54,010 passengers on the North Atlantic routes. As a result, the Belgian carrier expected capacity to increase from 1,596 in 1959 to 2,016 in 1960 on flights to Europe. In anticipation of the increase in demand for the 1960 peak season, the Belgian airline stated it would provide seating on its 707-329s for 142 passengers, 124 in economy class and 18 in first class. Sabena's non-peak configuration for economy class was 96, and 34 in first class. Furthermore, in 1959, Sabena's charter flights transported 3,296 passengers on the North Atlantic route network.

In September 1960, Sabena, with two DC-7C freighter conversions delivered in the fall, reported a 96% increase in its North Atlantic cargo service during the first half of the year. The airline accounted for 991,492 kg in the first six months of 1960, compared to 504,038 kg in the previous year. Cargo supply for the first half of 1960 stood at 1,420,651 kg, while in 1959 it was only 753,961 kg. Sabena's Cargo operations were showing solid signs of improvement.

Sabena opened two important routes in December 1960. The Belgian national carrier, which was struggling to overcome the economic impact of the loss of its routes in its former colony, the Belgian Congo, was busy looking for new avenues to fill its schedule. The first was introducing a new service to Mexico City, which Sabena and the Mexican aviation authorities agreed upon. The bilateral agreement stated that Sabena could fly to the Mexican capital via Montreal with two weekly flights. The new route to Mexico City was inaugurated on December 2, 1960, and was operated by Sabena's new Boeing 707-329 'Intercontinental' jetliners. The new route marked Sabena's first incursion into the Latin American market. Moreover, it was an intelligent move by the airline. It would help offset the revenue loss from its discontinued Congo routes, constituting approximately 40% of the Belgian carrier's annual revenue.

Regarding the Canary Islands, the Belgian airline entered into a pool agreement with Aviaco (Aviacion y Comercio). This small Spanish airline had the right to fly to Las Palmas, the capital of the Canary Islands. Under the plan, Sabena-owned Douglas DC-6 aircraft were painted in both Sabena and Aviaco colors for the flight from Brussels to Barcelona on Tuesdays and Saturdays and Madrid on Thursdays. Once in Spain, the aircraft was taken over by Aviaco aircrews and flown to Las Palmas. It was the first time in Europe that this form of cooperation took place, whereby crews of one airline would fly the first leg of a trip, and the crews of the second carrier would take over for the second leg. It was a win-win situation for both airlines. Sabena could fly its passengers to a destination with a strong tourist attraction, and Aviaco was provided with suitable aircraft to exploit its route to the Canary Islands effectively.

In the Near East, for the first time, Sabena's Brussels-Athens-Beirut-Tehran route was flown every Thursday by a Boeing 707-329 'Intercontinental.' That enabled Sabena to reduce the time of the flight from 13 hours 10 minutes to 6 hours 45 minutes. Meanwhile, the flight on the Brussels-Beirut route was also reduced from 8 hours 25 minutes to just 4 hours 5 minutes.

In the summer of 1960, Sabena carried 21% more passengers than the previous summer's 29,966. However, during the summer, the airline could not provide transatlantic services due to its participation in the Belgian Congo airlift of Belgians and other Europeans. Sabena flew 2,854 transatlantic passengers to New York between July 10 and 22. The peak weekly Brussels to New York capacity was 2,196 seats, up 38% from the rise in 1959. During the summer of 1960, Sabena operated six charter flights, all of which were on its jetliners.

On November 1, Sabena's only Cargo service winter timetable was released. Sabena's cargo route network, inaugurated four years earlier in 1960, served ten cities in five Western European countries from Brussels. They were as follows: France, Great Britain, the Netherlands, Italy, West Germany, and the US, which was the New York cargo service. Moreover, Sabena also used its passenger aircraft to carry cargo in their holds. The other benefit of Sabena's cargo services was that its cargo network provided Brussels-National airport with direct connections, without customs formalities, and 101 cargo destinations of its overall cargo network. The number of countries served by Sabena's cargo services was 37. In Europe: Brussels-London: seven weekly services (Sabena-BEA pool). Brussels-Paris: six weekly services. Brussels-Amsterdam: five weekly night services, each directly connected to Copenhagen, from Schiphol Airport. Brussels-Milan: three weekly flights, one weekly service via Nuremberg, with the return flight via Nuremberg and Dusseldorf. Brussels-Stuttgart: five weekly services; on the return flights, four flew back via Cologne, and one flight via Dusseldorf. Brussels-Nuremberg: One weekly service, with the return flight going via Dusseldorf. America: Brussels-New York, two weekly night services via Manchester, started operations on December 17, 1960.

The Congo Crisis and Sabena's Congo Airlift

At this point, I will describe the events which caused the Congo crisis in 1960. It is by no means a complete account of what transpired in the Congo in 1960, but it gives the reader some idea of what occurred during those terrible months in the Belgian Congo.

In early 1959, events that led to the battle for independence in the Congo started south of the colony, at Leopoldville. A Congolese ethnic political association, ABAKO, was the main instigator of riots in the capital city of Leopoldville. Joseph Kasavubu headed the association and was a vocal opponent of Belgian colonial rule. The Belgian government had no jurisdiction over most of the lower parts of the Congo, where ABAKO had a lot of support. However, the Belgian authorities in Leopoldville tried to prohibit ABAKO from assembling in large numbers in the capital of the southern province. That resulted in massive violence in the city from January 4 to 7, 1959, in which 34 indigenous Congolese were killed.

On January 12, 1959, Joseph Kasavubu was arrested. The Belgian authorities stated that he would not be released until March 13. However, the sources added that the result of the political process would eventually lead to elections, which would be held soon, and would lead to eventual independence for all of the Belgian Congo. That statement resulted in a surge of political activity, with over 50 political parties being registered, nearly all of them centered on tribal groups. As parties competed for votes, nationalist demands became more extreme. In response to growing insecurity, the Belgian Government convened a roundtable conference in Brussels for the leaders of the various Congolese factions. On January 27, 1960, after the meeting, it was agreed that elections would be held on May 22, 1960, and independence would be granted on June 30, 1960.

In the May 22 elections, only the two biggest parties presented themselves in more than one province. The 'Congolese National Movement,' or MNC-L, led by its left-wing leader, Patrice Lumumba, was the party that won the election, and Lumumba became President. At the same time, Joseph Kasavubu was the prime minister. The 'Parti National du Progres' or PNP was the other national party. The Belgian side favored it, but it came in second.

The situation in the Congo throughout June and early July was not calm, even though a Congolese leader led the Government for the first time. Then on July 5, the Congo crisis started when a mutiny broke out at Leopoldville (modern-day Kinshasa) by the Belgian created army, the 'Force Publique,' which was made up of many indigenous soldiers. They rebelled against their white Belgian leaders at the military camp of Thysville, demanding higher pay, more opportunities, and more control. The revolt spread swiftly to neighboring bases, and bloodshed erupted throughout the colony. It resulted in the subsequent intervention of Belgian paratroopers deployed in the capital to protect the white Belgian and other European residents' lives.

Thousands of Europeans, primarily Belgian, tried to escape from Congo as newspapers worldwide published accounts of white civilians, including women and children, being attacked, raped, or killed, and white-owned properties looted. The Belgians called in troops to restore order because they could not control the 'Congolese National Army.' They did not ask Prime Minister Kasavubu or President Lumumba for permission, incensed the newly formed Congolese army, and more soldiers joined Lumumba's group. The Congolese government requested soldiers from the United Nations and demanded that Belgian forces be removed. Two days before the outbreak of hostilities, the affluent Katanga province declared independence from the Republic of Congo in the far south of the colony. Belgian business interests dominated the Katanga province, and Belgium supported its secession from the Congo.

On July 13, the United Nations General Assembly responded to the crisis by passing a resolution authorizing the formation of an intervention force, the 'United Nations organizations in the Congo,' or 'ONUC' for short. The UN also passed a resolution demanding the departure of all Belgian forces from the Congo. Meanwhile, the killing of civilians continued, as did the killing between Congolese factions. Most Europeans equipped themselves with firearms while waiting to be airlifted from the hell that was the Congo. The European population of the Congo was terrified and distrustful of the indigenous people and leaders. Many European settlers determined to stay in the Congo three months earlier were willing to give up their African homes and leave the Congo, never to return. The power of authority had firmly moved to local political leaders in several parts of the Congo. Even they, however, were unable to sustain power in the face of other Congolese political opponents with whom they were locked in a massive conflict.

Sabena entered the conflict on Saturday, July 9. Belgian aircraft landed in the Congo and took the first load of refugees from the inferno of the colony to the safety of their ancestral homes. It was not the first time Sabena had participated in a significant airlift operation. In August 1950, it supported the American operations in the Korean War Crisis with its Douglas DC-4s. This airlift, however, was much closer to home. The management of the Belgian national airline stopped all scheduled operations. It focused on evacuating the Belgian and other non-indigenous people out of the Congo as expeditiously as possible. Sabena's 'Air Bridge' would save many lives. Its air bridge covered 3,730 miles (6,000 km), and the average flying time from Belgium to the Congo was 15 hours. The number of Sabena aircraft taking part in the evacuation of the Congo was 44. The rescue fleet consisted of 5 Boeing 707s, 8 DC-7s, and 12 DC-6s, as well as a further 19 aircraft from Sabena's Congolese network. The Belgian carrier's squadron of aircraft was led by its five invincible Boeing 707-329s, which shouldered the most significant number of duties during the crisis. From the beginning of the operation on Thursday, July 7, no aircraft from the fleet of 44 airliners parked at an airport for a lengthy time, except for necessary maintenance and repairs. The plane took off soon after they dispatched the refugees in Belgium and headed to the Congo at breakneck speed to save the next lot of evacuees.

In the first week of the Airlift, Sabena's aircraft evacuated 8,428 people. The carrier could have taken many others out of the Congo in that first week if it had more aircraft. Fortunately, help was offered to Sabena by many European airlines, which soon sent aircraft from their fleets to save the people in the Congo. From Paris, Munich, Amsterdam, Rome, London, and several other European cities, planes took off for the Congo to shore up Sabena's airlift. Some other airlines serving destinations in Africa interrupted their scheduled flights and stopped in Brazzaville and Leopoldville to fill any available seats on the aircraft with evacuees from the Congo. The pilots and crews from Belgium, and the many other European countries that assisted the Belgian nation in its hour of need, doggedly maintained the airlift so as not to allow the innocent people in the Congo to suffer the horrors of the tragic events unfolding there.

On an average day during the crisis, the five Sabena 707 jetliners and the airline's seven DC-7Cs and eight DC-6 propliners flew out of the Congo, transporting 2,500 passengers. Moreover, depending on the weather, a Sabena 707 flew one round-trip per day between Brussels and the Congo, covering a total distance of 7,738 miles (12,450 km) in 14.5 to 16 hours, a phenomenal feat by anyone's standards.

The airlift from the Congo lasted until July 28. During that time, 166 Sabena pilots were designated for the flights. Together, they covered over 1.86 million miles (2.9 million km) in just 19 days and evacuated 25,711 people, of whom 1,888 were infants, 8,327 children, 9,742 women, and 5,754 men. Sabena achieved this total in 209 flights. The Boeing 707-329s made 62 flights, the DC-7Cs 66 flights, and the DC-6/DC-6Bs completed 81 flights.

First among the airliners were the 5 Boeing 707-329s. Sabena's Boeing 707s, which generally carried 170 economy class passengers, were stripped of their armrests, creating more space between the seats to accommodate up to 300 passengers. One event that stood out, one such 707, broke a record in the number of people it brought back to Belgium. On July 17, the Boeing 707 jetliner carried 303 women and children from Leopoldville to Brussels. That example highlighted the dependability and durability of the new jetliners. Moreover, it emphasized that civil aviation could play an important humanitarian role when exceptional circumstances demand it.

The pilots, the navigators, the cabin crew, the mechanics, the ground staff, and the office staff of Belgium's national carrier all worked as one. Their passion and urgency were to save the lives of people they did not know and would probably never see again once the operation ended. They knew that they were ordained to do what they had to do because human life was sacrosanct. They worked day and night with sweat, fear, tears, emotion, and joy with little sleep and minimum comfort. That did not matter to them, for saving a life in danger is one of the most extraordinary tasks one could experience, 'whoever saves a life saves the world.' Praise must be given to those gloriously brave souls of Sabena who assisted the innocents in the Congo, caught in the middle of an abyss, evaded darkness, and led them to safety. The intense emotions released by Sabena's staff in those days during the Congo airlift provided the impetus for completing the evacuation of the people from the Congo quickly and successfully. The people of Belgium will never forget the brave efforts of all the employees of Sabena during the tragic events of the Congo Crisis.

Photo 39 (left) & photo 40 (right) Evacuees from the Congo smile as they return home to Belgium in one of Sabena's Douglas propliners. (Photos 39 & 40 by Herbert Behrens of Anefo, via Nationaal Archief The Netherlands)

In the aftermath of the Congo crisis, Sabena's management saw a major problem looming. Sabena was almost certain to face substantial revenue loss stemming from the carrier's emergency Congo refugee airlift that took away its aircraft from its most lucrative markets. Before the Congo Crisis, the Belgian carrier served 43 African cities and towns, 37 of which were in its former African dominion, across a network of around 51,200 miles (82,400 km). Moreover, Sabena routes connecting Brussels and Africa totaled 136,000 miles (218,870 km). A significant chunk of Sabena's revenue was obtained from its Brussels to Congo routes and its Congo domestic route system, which accounted for 40% of its overall revenue. The possibility of irreversible loss of those routes was a substantial concern for top Sabena officials. Sabena's management understood that if it had to flee the Congo, it would be replaced by many other Western carriers already operating within the African continent, such as BOAC, Air France, and KLM.

Sabena had committed its whole fleet of airliners to the emergency operation between July 9 and July 28, when some airlift planes resumed commercial service. Within the Congo, the carrier used its four DC-4s, four Convair CV-440 Metropolitans, and 13 DC-3/C-47s to transport the evacuees from the interior to significant evacuation sites like Leopoldville and Elisabethville, as well as the Belgian military base of Kamina in central Congo.

Before the crisis, on its long haul services, Sabena had been operating its 5 Boeing 707s on weekly schedules to the following destinations: Africa, with eight flights to Leopoldville, of which two went via Elisabethville, and two via Johannesburg. In Eastern Europe, there were three flights to Moscow, and in North America, it provided 13 weekly flights to New York. New York was the biggest concern for management. Amid the Congo upheaval, Sabena lost almost BFR 25,036,023 in potential earnings on the important Brussels to New York route. Sabena's propliners, the DC-7Cs, and DC-6/6Bs utilized in the airlift were also taken out of regular service in Europe, the USSR, Athens, Ankara, Cairo, and Teheran and Leopoldville, which meant further losses for the Belgian airline.

Photo 41. It might seem to be a mixed Sabena & Congo airline, 'SABENGO,' but it is Sabena 707-329 'OO-SJC' delivered to the Belgian airline on February 13, 1960. The mixed titles are due to the airliner being leased to Air Congo several times between 1962 and 1967. This photo was taken in May 1966; thus, the aircraft must have been in service with Air Congo. (Photo by John F. Ciesla)

1961

In early January, the Congolese Government, led by Prime Minister Kasavubu, and Sabena, reached a protocol agreement to create a Congolese national airline company named Air Congo. The Congolese government would control the majority of that company. However, the Congolese Government was in turmoil in the aftermath of the Congo Crisis, causing delays in the decision to form the new airline. The Congolese Government and Sabena signed the agreement just months after the Congo Airlift. Since that time, Sabena had made remarkable efforts to maintain the continuity of its air services in the domestic Congolese network after the independence of the Congo. Eventually, the new national airline of the Congo was formed on June 6, 1961. On June 28, the airline was incorporated, with ownership divided between the Congolese Government 65%, Sabena's subsidiary Sobelair 5%, Air Brousse (a small privately-owned airline formed in the Congo in 1955) 5%, and Sabena 30%. The CEO of the new airline was Congolese Hubert Sangara, and the Vice President was Sabena's President, Gilbert Perier. Sabena succeeded in setting up Air Congo with the Congolese Government, safeguarding its Brussels-Leopoldville route. Air Congo was based in the capital Leopoldville. Sabena was chosen as the technical advisor to Air Congo. On June 29, the Congolese airline began operations with exclusive rights to all scheduled domestic Congolese and international flights. Air Congo was also given the right to apply for government subsidies for any incurred deficit.

The first Sabena Caravelle VI was handed over to the pilots of Sabena, who had been training on the type at Toulouse, on January 2, 1961. The second jetliner for Sabena was now ready. Comparing the Caravelle VI to the original Caravelle III model, the former was powered by two Rolls Royce Avon Ra-29-531 turbojets. They provided 5,670 kg thrust and were more potent than the Caravelle IIIs Avon Ra-29-527 engines, which delivered 5,310 kg thrust. Its cruising speed was 511 mph (822 kph) at 30,000 feet. It had a range of 1,410 miles (2,270 km), and the Caravelle VI flew almost 300 miles further than the Caravelle III, which had a maximum range of 1,120 miles (1,800 km). The Caravelle VI was also a heavier aircraft, with a takeoff weight (TOW) of 48 tons, compared to the 46 tons TOW for its older sibling. Concerning accommodation, the VI provided seating for 16-20 first-class and 55-60 economy class passengers or 80 economy class passengers. Sabena was the launch customer for the new Caravelle VI, which made its first flight on September 10, 1960. Sabena intended to use its Caravelle VIs intensively. From April 1 until October 31, the Belgian carrier's new Caravelles were expected to make 4,720 take-offs and landings for the summer timetable period. Operating in a network with 22 destinations in Europe and the Near East, which had stretched to 12,430 miles (20,000 km), Sabena's new 6 Caravelle VIs would have a busy summer. The Near East had become an essential region for Sabena. The Belgian company planned to gradually extend its Caravelle network inward towards the Middle East, with all flights from Brussels passing via Athens.

Sabena opened its Caravelle service on February 18, 1961. By the end of August, the French jetliners had made 1,786 flights, flew 4,818 hours, and transported some 90,000 passengers. Sabena was so impressed with its Caravelles that it ordered two more Caravelle VIs.

Sabena's operating statistics for 1960 showed that the carrier achieved a considerable increase in traffic performance compared to 1959, even with the events in the Congo. Passenger traffic rose dramatically by as much as 45%. Additionally, in 1960 the ton-km offered and utilized numbers reached the highest amount since 1956, and the number of cargo ton-km also increased prominently in 1960. On the other hand, the number of mail ton-km dropped the previous year due to the Congo Crisis, vastly reducing postal traffic between Belgium and its former overseas possessions. Why the vast rises? The increases were primarily attributed to the high productivity of the Boeing 707-329 jetliners, which could carry about three times the number of passengers compared to the propliners. The second reason was that Sabena always had high load factors, averaging almost 80% each year, on the Belgium-Congo routes and the Congo domestic network, due to the nearly total domination in its colony right up to the Congo Crisis. Those load factors accounted for around 40% of Sabena's overall traffic. The final reason was Sabena's ability to reduce its cost level by 10%. The Belgian airline achieved that because it increased its capacity by 38%. Of course, there were casualties; however, to get the increase, Sabena had to reduce its workforce by 2% to 10,500. Most of the personnel that lost their jobs were the Congolese staff. At the same time, Sabena kept tight control on its home-based Belgian employees.

Sabena's helicopter service on the Brussels-Paris route took a hit in 1960. Although the flights provided by Sabena in the previous year were on a regular basis, the number of flights offered fell sharply, resulting in a drop of 19% in passenger traffic. In the summer of 1961, Sabena's helicopters served the following routes: Brussels-Antwerp-Rotterdam with 18 weekly flights. The number of flights was also 18 per week on the Brussels-Liege-Maastricht-Cologne line. Six of those flights extended the service to Bonn. At each of the following destinations, the frequency of flights was 12 per week: Lille, Duisburg, Dortmund, and Eindhoven (only cargo flights). The Brussels-Paris route, which saw a drop in passenger traffic in 1960, recovered somewhat in 1961. Of the 40 flights that Sabena provided to Paris in 1961, 27 were by its airliners and 13 by its helicopters.

The summer timetable for Sabena's regular cargo network came into effect on April 1, 1961. Sabena's cargo network, which began operations five years earlier, offered cargo services from Brussels to 10 Western European cities, which were: Paris, London, Amsterdam, Eindhoven, Milan, Cologne, Dusseldorf, Nuremberg, Stuttgart, and Vienna. One of the cities in the cargo network, Eindhoven, was serviced by Sabena's helicopters. The route was flown five times a week by a Sikorsky S.58, which could carry a maximum load of approximately 1,100 kg. In addition, Sabena placed a DC-7CF freighter on that line on the New York cargo route from February 11. Sabena's long-haul cargo services to the United States, Canada, Central Africa, and South Africa operated without customs formalities at Brussels Airport, making things easier for the airline.

In July, Sabena ordered two additional Boeing 707-329s with the same type of engines as the four 707s already in service with Sabena, the Pratt & Whitney JT4A-3 turbojets. Once delivered, the 707s were scheduled to be put into service on the Sabena's international routes to the United States, Canada, Mexico, and Africa. The two new 707-329 jetliners were delivered in the spring of 1962.

The annual general meeting of shareholders of Sabena was held in July. It provided an opportunity to take stock of the development of the Belgian national company, especially after the events of the Congo in 1960. Therefore, the balance sheet showed a deficit of more than BFR 131.7 million. A repayable advance from the state covered the deficit, which allowed certain amortizations to be made under Sabena's statutes. Sabena's results for 1960 were also burdened by the costs of activating its Boeing 707-329 and Caravelle VI jetliners. The arrival of the jetliners led to Sabena replacing its piston-engine airliners. Additionally, the Governing Body's report emphasized that the results would have been much better had Sabena not used its entire fleet of long-haul aircraft, and its entire fleet based in the Congo, exclusively on the Congo Airlift between July 9 and 28. This mobilization of almost all of the Belgian carrier's aircraft severely disrupted seasonal traffic right in the middle of the high season. Thus, revenue per ton-km went from BFR 21.09 in 1959 to BFR 19.61 in 1960, a reduction of 7.02%. Concerning Sabena's performance in 1960, the carrier recorded a significant increase in its traffic figures. The supply of ton-km increased by 30%, from 217,947,098 in 1959 to 300,773,194 in 1960. Moreover, the ton-km used increased by 45%, from 152,191,525 in 1959 to 220,902,281 in the previous year. Excluding the transport services provided during the airlift, Sabena's growth rate for overall transport services in 1960 was 22%. The number of passengers carried in 1960 was 928,683, compared to 730,576 in 1959, up by 26%. Despite fewer flights, Sabena transported 33,689 people in 1960, a 3.6% increase over 1959 on its helicopter network. That was due to a systematic reduction of the helicopter fleet in anticipation of an overhaul of the whole of Sabena's helicopter network. In the cargo division, services increased by 20%. Whereas in the mail division, there was a significant decline of 25% following the events in Congo. At the end of 1960, Sabena served 96 stops in 39 countries. Its staff numbered 10,506. Its fleet included 5 Boeing 707-329, 8 Caravelle-VI, 8 DC-7C, 10 DC-6B/DC-6, 9 Convair CV-440 Metropolitans, 28 DC-3/C-47, 6 S.58 helicopters, 1 Bell 47 helicopter, 1 Alouette-II helicopter, and 22 training aircraft. It had two 707-329s on order.

Sabena's winter timetable for 1961 featured some significant improvements. First, Sabena's route network served by jet aircraft was rapidly expanding. Caravelle VI jetliners serviced Rome and Istanbul. On other network routes, the number of flights increased compared to the 1960 winter season. For example, the Middle East had more flights than the previous winter. On the transatlantic network to New York, Montreal, and Mexico City, there were 11 weekly flights. Nine of those flights were on the Boeing 707-329 jetliners, and two flights were with the Douglas DC-7C 'Seven Seas' propliners. With the reintroduction of the DC-7Cs on the Atlantic route, Sabena could carry out flights in an all-economy class, allowing the airline to offer its customers surcharge-free jet flights. While the number of flights on European routes was traditionally reduced during winter seasons, weekly flight frequencies were higher in the 1961 winter season compared to the winter of 1959/1960. From November 1, 1961, to March 31, 1962, Sabena operated from 222 to 249 weekly flights. More than 50 flights were flown by the carrier's 707, or Caravelle jetliners. They were the transatlantic flights and flights to Africa and the Middle East. Sabena continued to operate the route to the Canary Islands in cooperation with the Spanish airline 'Aviaco.' That route followed the Brussels to Tenerife-Las Palmas via the Madrid route. Then from September 30, Sabena also offered a flight to the Canary Islands with a stopover at Lisbon.

The end of 1961 was not a good time for Sabena. In December, the Belgian airline revealed that it had made a loss in the order of BFR 600 million. Sabena suffered greatly from the events in the Congo, which seriously affected the flights on its vital Brussels-Leopoldville line. It was also profoundly affected by the negative results on the North Atlantic network, where Sabena's average load factor did not pass 42% during the peak of the summer period. While the airline's capacity increased by 53% compared to 1960, its air traffic increased by only 3%.

Photo 42. Sabena DC-6B 'OO-CTN' at Rotterdam (EHRD) in 1970. The aircraft was delivered to Sabena on February 25, 1954. It was later leased to Caledonian Airways in April 1964 and was returned to Sabena in November 1964. In September 1971, it was sold to Gabonair.
(Photo from 'www.douglasdc8.com')

1962

The year 1961 was Sabena's second year of jet operations. The first year in 1960, Sabena had only the 707s at its disposal. Then, on February 8, 1961, the first two of eight Caravelle VIs ordered from Sud Aviation were delivered to Sabena in Brussels. Just ten days later, on February 18, the first of two Caravelles received by Sabena was placed in regular service between Brussels and Nice. By August 1961, all eight Caravelles had joined Sabena's fleet. When all Caravelles were delivered in August, Sabena's medium-haul network had significantly expanded. Sabena's Caravelles carried out 4,165 flights and completed approximately 14,300 takeoffs and landings in their first year of operation. In addition, they covered around 3.6 million miles (5.8 million km) and logged 12,443 hours of flight time. Some 214,960 of the 562,000 passengers transported by Sabena from 1961 to 1962 on its medium-haul network flew on its Caravelles. Furthermore, the Belgian airline's French super jets carried 1,471 tons of cargo. Therefore, it was no wonder that Sabena had a full summer schedule set up for its Caravelles. The carrier planned to make 4,353 scheduled flights with its Caravelles in the summer of 1962.

On the topic of Caravelles, on January 8, 1962, an international incident involved Belgium and the USSR. A Sabena Caravelle with 27 people on board, flying en route from Tehran to Istanbul, was detained at a Russian military airport near Yerevan, in Russian Armenia, after Soviet fighter jets forced it to land when it had flown off course. At 4:30 am, not long after taking off from Tehran, the aircraft's instruments stopped working, and the pilot radioed for assistance. At 6:20 am, the Caravelle pilot sent a second radio message, saying that his Caravelle was intercepted by Russian fighter jets and forced to land. He added that he followed the Migs and flew over mountainous terrain. He claimed his airliner flew above Yerevan, the capital of Soviet Armenia. After that message, nothing further was heard from the pilot. Later, Sabena officials revealed that the airliner's radio compass was responsible for the Caravelle drifting off course. The following day, cool heads prevailed, and the international USSR versus Belgium affair fizzled out. All 18 passengers on the flight and half the crew members were flown back to Brussels by a Tupolev jetliner via Amsterdam. However, the Soviet authorities kept Sabena's Caravelle at Moscow airport, with four of its crew members, who did not fly back to Brussels. Eventually, the two nations worked out the problem, and the four crew members, together with the Caravelle, were returned to Belgium.

Sabena was not only a large international airline with many destinations worldwide, but it also operated an air taxi service within Belgium and its neighboring countries. The least known aircraft in Sabena's repertoire were just as important to the carrier as any of its large aircraft. Sabena had eight twin-engine Cessna 310s, which could carry five people, and were used primarily for training the carrier's pilots. They were also air taxis and could be booked by tourists and business people to fly to their desired destination. In 1961 alone, the Sabena air taxis made 361 flights to about 30 European cities, half of which were not connected on the regular European air networks. In 1961, each flight by Sabena's fleet of small aircraft averaged 68 minutes, and for the whole of that year, the Cessnas flew 47,550 miles (76,524 km). Moreover, they transported around 1,200 passengers with an average distance of 132 miles (212 km). The destinations chosen by the passengers were the following cities: Paris, Strasbourg, Dijon, Dole and Toulouse in France, Hanover and Braunschweig in Germany, Turin in Italy, Basel Lausanne in Switzerland, Bilbao and San Sebastian in Spain, Oslo in Norway, and Birmingham in Great Britain. Not a bad day's work for the spirited Cessnas in Sabena's service. In April 1962, Sabena announced an order for eight more Cessna 310s.

Sabena's fleet in early 1962 consisted of the following aircraft: 4 707-329, 8 DC-7C, 7 DC-6B, 8 Caravelle 6, 2 DC-6, 9 Convair CV-440, 7 DC-4, 29 DC-3. Helicopters: 5 S-58, 1 Alouette II, 1 Bell 47H-1, Light aircraft: 8 Cessna 310. It had 2 Boeing 707-329s on order.

Sabena announced on June 5 that 22 additional flights would be made available to Usumbura, the capital of Rwanda-Urundi, from June 10 to July 1. Those flights were related to the fact that the small Belgian colony bordering the Congo was to become independent on July 1. In addition, the Belgian Government had soldiers stationed at Rwanda-Urundi to ensure a swift and safe exit from the small colony, unlike what had resulted in the Congo in 1960.

The Belgian airline had a disastrous year in 1961, as it incurred significant losses. The board of directors stated that the carrier's financial results for the 1961 financial year were affected by a substantial decrease in the utilization rate, which was the lowest since 1949, causing the airline to lose profitability. Oversupply on the North Atlantic route contributed to the reduction in load factor, which fell from 73.4% in 1960 to 63.3% in 1961. Sabena placed its largest airliners on the North Atlantic route network due to their larger capacity, as it expected a multitude of tourists flying to the old continent. However, Sabena's expectations did not pay off. A corresponding drop in demand could not offset the increased capacity, as many Americans did not travel to Europe in 1961. Many Europeans surmised that the decline in flights was due to factors of economic and political nature which forced most Americans to not travel to Europe in 1961. Thus, Sabena was stuck with half-filled airliners flying across the Atlantic. It was not only Sabena that was affected; lower demand affected all major airlines serving the North Atlantic route in 1961 to varying degrees. The other factor that caused Sabena's problems was the obvious one, the Congo. In 1960, the 'Congo Airlift' set up by Sabena greatly affected the results of that year, especially as the airlift took place at the worst time of the year for the Belgian carrier, the summer season. Moreover, Air-Congo's new airline, founded in the Congo in 1961, took over Sabena's Congolese domestic network on July 1, 1961. Those were the reasons that left Sabena's management feeling despondent in 1962.

Photo 43. Cessna 310B 'OO-SED' ex Sabena trainer. Delivered to the airline in 1957 and was operational until 1982. It is pictured here still wearing Sabena colors, with US registration 'N611CE', at Cranfield, UK, in 1985. (Photo by Andy Chetwyn)

Photo 44. Cessna 310B 'OO-SEG' was with Sabena from 1958 to 1982. (Photo by Richard Vandervord)

The airline's annual report showed that all expenses and charges in 1961 were BFR 4.374 million compared to 4.394 million in 1960. Operating income and other income in 1961 amounted to BFR 3.985 billion compared to BFR 4.263 billion in 1960. Also, compared to 1960, expenditure and charges fell by 0.5% with a 2.5% drop in production, while revenue was 10.7% lower following a 16% drop in traffic.

One positive for Sabena in 1961 was its freight division, which posted a handy increase of 19.7% in ton-km for 1961. The airline's cargo services seemed to be unsusceptible compared to Sabena's passenger services which took a beating the previous year. From November 1, Sabena's C-47 and DC-7C cargo planes provided freight services to 1,165 European routes and 86 transatlantic ones. Those flights amounted to a capacity of 6 million ton-km and a load of 4,000 tons. Sabena's cargo services began operations six years earlier, and they were going strong. The cargo flights originated from Brussels and flew to New York on the transatlantic network, with one service flying via Montreal and the other via Manchester. On the Brussels-Paris route, there were six weekly night cargo services. Also, from November 1, Sabena provided 27 weekly flights from Brussels to Paris on its passenger services, some of which were flown by the Caravelle VIs. The Caravelles were also used on the Brussels-Nice line six times a week until February 1963, when they increased to 8 times. Rome was also in the picture, with a total of 19 stopovers by Sabena's Caravelle VI jetliners.

Sabena received some excellent news in mid-November. First, Sabena signed a general agreement with Air Congo. From January 1, 1963, the contract stated that a weekly nonstop flight, Brussels-Leopoldville-Brussels, would be provided and a weekly direct service Brussels-Rome-Leopoldville-Rome-Brussels. The flights would be flown by Air Congo, using Sabena's Boeing 707-329s. They would represent around 20% of the traffic between Belgium and the Congo and be managed by Sabena's technical staff. In addition, at the end of December, Sabena provided two Douglas DC-3s and two Douglas DC-4s to Air Congo and the corresponding batches of reserve engines, accessories, and spare parts. That roughly represented 27% of the fleet based in Leopoldville and assigned to Air Congo's domestic network.

The end had finally arrived for Sabena's helicopter services. In November, the Belgian carrier stated that it planned to end all its helicopter operations soon. Strong protests met that declaration, and the company decided to leave two Sikorsky S.58 flying, albeit on a reduced network linking Brussels to six destinations, for the forthcoming summer season. The network in November only included four routes: Brussels-Duisburg: two weekly services via Cologne with optional stopovers in Liege and Maestricht; Brussels-Cologne: six weekly services with optional stopovers in Liege and Maestricht; Brussels-Rotterdam: six weekly services, 12 from February 4, 1963, all flying via Eindhoven; and the Brussels-Eindhoven route with 12 weekly services. The Brussels-Paris service had already been abolished. That route was now serviced mainly by Sabena's Caravelle VIs.

1963

The year 1963 was the 40th year of Sabena's existence. The Belgian carrier, which was one of the oldest airlines in the world, had flown more than 310 million miles (500 million km) since its inception in 1923. More than half of the 9 million passengers it had carried since its operations began were transported in just the previous five years. In 1962 alone, Sabena moved an average of 2,540 passengers per day on all its flights, while back in 1924, it had transported only 1,789 for the whole year. Also, in 1924, the average capacity of Sabena's aircraft was only 600 kg. In 1963, it was a substantial 8.5 tons. Regarding the mail, and cargo numbers; in 1962, the number of ton-km of mail carried by Sabena was 1,170 times that of 1924, while the number of ton-km of cargo was 3,360 times greater.

Photo 45. Sabena Caravelle VI 'OO-SRH' at Brussels in May 1966. Delivered on August 11, 1961.
It remained with Sabena until sold to Aerospatiale on September 3, 1973. (Photo by John F. Ciesla)

Photo 46. Sabena's fifth Caravelle VI, 'OO-SRE, ' seen here in Brussels in May 1966. Delivered on March 24, 1961.
Sold to the French airline Catair on June 3, 1976. (Photo by John F. Ciesla)

Photo 47. Another shot of Sabena Caravelle VI, 'OO-SRE,' at Zurich Airport on July 8, 1967. (Photo from ETH Zurich)

Photo 48. Sabena Caravelle 'OO-SRH'. On the tarmac at Zurich on July 6, 1970. (Photo from ETH Zurich)

Photo 49. Sabena Caravelle 'OO-SRF' at Munich Airport in September 1976. Delivered to Sabena on April 28 , 1961. Sold to EAS - Europe Aero Service on June 5, 1978 (Photo by Norbert Kröpfl)

Photo 50. Sabena Sud Aviation SE.210 Caravelle 6 'OO-SRC' at Liverpool John Lennon Airport on November 26, 1964. This was Sabena's third Caravelle, delivered on February 16, 1961. Sobelair leased it from April 1971 till the end of 1975. It was finally sold to Europe Aero Service with reg. 'F-BYCA. ' (Photo by Ken Fielding via Wikimedia)

In terms of aircraft performance, in 1923, Sabena's largest airliner was the Handley Page W8b, with accommodation for ten passengers and a speed of 75 mph (120 kph). In 1963, its Boeing 707-300 'Intercontinental' jetliners provided accommodation for 164 passengers and flew at 520 mph (840 kph). In 1963, the carrier's eight medium-haul Caravelle VI jetliners and six Boeing 707-329s accounted for 83% of its total production. During its 40 years of existence, Sabena had used 41 different types of aircraft, and in 1963 the Belgian carrier stood at the forefront of technical progress. In the three years since the introduction of jetliners to its fleet, Sabena's turbojets carried 400,000 passengers and flew 4,605 flights, corresponding to 42,568 flight hours.

Sabena made significant changes to the Belgium-Congo line on February 13, 1963. Sabena expanded one of its four weekly flights to the Congo after the Elisabethville (modern-day Lubumbashi) Airport reopened following renovations that allowed jetliners to land and take off. As a result, the line between Brussels and Athens was extended to Elisabethville, the capital of Katanga in southern Congo. The Belgian carrier's DC-7Cs would operate the other three weekly flights. The routes for the DC-7Cs were as follows: Brussels-Usumbura would have three weekly flights, while 2 of those flights would continue to the second-largest city in the Congo, Elisabethville. As for the Congolese airline, Air Congo was to provide weekly services with its leased Boeing 707s from Brussels to Leopoldville (modern-day Kinshasa). The flight would leave Brussels every Monday and return to Leopoldville every Thursday. The route would be expanded to twice weekly beginning in April 1963. The summer timetable showed that Sabena's network had grown slightly in April.

In Europe, Sabena added a new city to its network, Hanover. Previously, the German city had only been served by Sabena's airliners in April and May when that city hosted some international exhibitions there. Concerning the rest of Germany, throughout the entire 1963 summer season, Sabena's aircraft would provide three weekly flights to Hanover, only on weekends. Meanwhile, Frankfurt would benefit from 3 flights which would be via Luxembourg. Sabena also offered daily flights to Stuttgart, Munich, and Hamburg.

Additionally, from 1963, the Brussels-Hamburg line would be nonstop. Six weekly flights were penciled in between Brussels and London for the UK. Sabena's Caravelles flew 3 of those flights, while BEA operated the other 3. In summer, there would be four daily connections between Brussels, Paris, and Amsterdam. The flights would be operated jointly by Sabena, SAS, and KLM on the Amsterdam line. There were also two daily flights to Copenhagen and Stockholm, also jointly operated by Sabena, SAS, and KLM. Switzerland was the recipient of 2 flights to Zurich, and 2 to Geneva, jointly run by Swissair and Sabena. The Italian city of Milan got itself two daily flights, and the capital Rome had 14 weekly flights, all jointly run by Sabena and Alitalia. Initially, Sabena would fly the Brussels-Nice line daily in the summer; then, from the end of June to the end of September, flights would be increased to 10 per week. Athens had daily flights from Brussels, which expanded to about 11 per week from May to September. In Austria, Vienna was serviced five times a week. Meanwhile, there were seven weekly flights to Madrid, six to Barcelona, two to Malaga, and five to Las Palmas for Spain. The flights were operated jointly by Sabena with the Spanish company Aviaco. Two weekly flights were also made available on the Lisbon and Casablanca routes. In Eastern Europe, Sabena provided regular services to Prague, Warsaw, Moscow, Budapest, Bucharest, and Belgrade. Sabena's Caravelles flew to Beirut, Istanbul, Cairo, and Tehran twice a week on the Middle East route network. On the Cairo route, there was an extra weekly flight with Sabena's DC-7Cs. There were also weekly flights to Israel.

On Sabena's long-haul network, transatlantic flights operated by Sabena's 707-329s increased from 11 to 17, with nine to fifteen departures each week to New York and two to four departures to Montreal. Sabena's operated its route from Montreal to Mexico with the Mexican national airline

Aeronaves de Mexico with two weekly flights. In summer, there were six weekly flights to Leopoldville via Geneva, Rome, Madrid, Athens, and Cairo and some nonstop flights in the Congo. Two of the flights were operated in cooperation with Congo's airline, Air Congo. There were also five weekly flights to the Katanga region's capital, Elisabethville, Boeing 707s or Douglas DC-7Cs via Geneva, Rome, Cairo, and Athens. Three of the Elisabethville flights flew via Usumbura. In addition, Sabena provided the South African city of Johannesburg with two weekly flights. In early 1963, Sabena had the following aircraft in its fleet: 5 Boeing 707-329, 6 DC-7C, 1 DC-7CF, 6 DC-6B, 8 Caravelle VI, 1 DC-6, 8 Convair CV-440, 2 DC-4, and 17 DC-3. Helicopters: 2 S-58, 1 Alouette II, and 1 Bell 47H-1. Light Aircraft: 4 Cessna 310. It had a Boeing 707-329 on order.

In August 1963, Sabena, which had initially decided to abandon its helicopter services, reversed its decision following a request from the Belgian Ministry of Communications to keep some of its helicopter services going. However, since the carrier was receiving subsidies from the Belgian Government, it had no choice but to continue to do so. Consequently, it kept the following routes in operation: The Brussels-Liege-Maastricht-Cologne tri-weekly line and the Brussels-Eindhoven-Rotterdam line with twelve flights a week.

Later in the year, Sabena and Austrian Airlines entered into a pool agreement on November 1, 1963. The agreement stipulated that the two carriers would cooperate on the Vienna-Brussels route, serviced by one of Sabena's Caravelles. The cabin crew would be Belgian and Austrian and would alternate. The two airlines would equally split the revenues and profits (or losses). The two companies agreed to start with five flights per week. After a bad year in 1962, in which the Belgian carrier's turnover was BFR 3.765 million and a deficit of BFR 470 million, the Belgian Government was obligated to pay a subsidy to Sabena by the law and Article 33 of Sabena's statutes. That enabled the airline to balance its books on its 40th anniversary without showing a loss. However, despite various economic measures implemented, 1962 was a considerably worse one than 1961, when the Government intervention amounted to BFR 389 million. The main reasons for Sabena's disastrous year in 1962 were the loss of most of its Congo routes and the carrier's increasing difficulty in securing traffic rights.

1964

In February, Sabena announced that it would suspend flights between Manchester and New York from April 1, 1964, following a decision by the British Ministry of aviation to stop most of Sabena's flights on the route and limit the Belgian carrier to only two weekly flights in summer. That decision put Sabena in an unfavorable position with BOAC, which operated daily jetliner services between Manchester and New York. According to the British Ministry of Aviation, the reduction in transatlantic flights via Manchester by Sabena in the summer was due to an aviation agreement concluded in September between Great Britain and Belgium. Sabena was the first airline to open a regular service between Manchester and New York in 1953 before BOAC appropriated the route.

In 1963, Sabena's performance results were satisfactory. That year Sabena carried more than 1 million passengers in one year for the first time in its history. In 1963, Sabena had almost reached the 1 million level but fell short with 992,000 passengers transported. After 41 years of operations, the Belgian airline had transported more than 10,500,000 passengers with its aircraft, more than half of that total within the previous five years. In comparison, it took the airline 27 years, from 1924 to 1950, to get to its first one-millionth passenger. In addition, Sabena reported an increase in supply in 1963, which was matched by the rise in ton-km sold, and attaining a reasonable load factor of 61.4%.

Moreover, the Belgian airline carried 3,906 tons of cargo on the North Atlantic network, increasing 31% compared to 1962. Mail transport also increased by 2%, and overall cargo tonnage rose by

23%. As a result, the carrier's annual deficit was not as bad as the deficit in 1962. Another exciting milestone for Sabena in 1963 saw the airline operate its 9,000th regular flight between Belgium and the Congo since operations on the African route began in 1935. Moreover, in 1963, Sabena's 56 aircraft in the fleet achieved 88,757 flight hours on its network, which had extended to 22,684,775 miles (36,507,608 km).

Sabena reached another '1 million mark' milestone in June 1964. The Caravelles carried their one-millionth passenger. Sabena's fleet of 8 Caravelle VIs had been in service since February 1961, and it took them almost 3.5 years to fly their one-millionth passenger. One can only show admiration for the legendary jetliner from Sud-Aviation. Sabena was a fan of the Caravelle. The carrier published a report summarizing the three years of service of its Caravelles. The report stated that Sabena's passengers since 1961 flew in 16,500 flights, with 44,000 hours. Furthermore, 7,000 tons of cargo were carried by the Caravelles in those three and a half years. One other startling fact was in Sabena's air traffic results. The Belgian airline's Caravelles accounted for 57.4% of all European and Middle Eastern route network flights. Caravelles flew to 30 of the 46 destinations on those two networks, an astounding statistic. Additionally, in their years of service, the French jetliners had offered 1.26 billion seat-km and produced almost 770.4 million passenger-km, resulting in a profitability factor of 61.1%. The Caravelles were the most robust jetliners in Sabena's fleet in the 1960s. Later in August, Sabena decided to reinforce its fleet of its Caravelle VIs when it ordered two additional types, which would bring the total number of Caravelle's in Sabena's fleet to 10. Delivery for the two Caravelles was scheduled for January and April 1965.

Sabena's summer timetable in 1964 did not bring any significant changes in Europe. There was the Brussels-Manchester route, which, for the first time, would be operated by the carrier's Caravelle VIs. The two summer seasonal routes, Brussels to Lourdes and Naples, were made available in 1964, as in previous summers. Two weekly flights were offered to Lourdes from May 2, increasing to 5 per week from May 26. The flights to Naples would begin on April 1, with 2 per week. Another seasonally operated route, which connected Ostend-Middeikerke (a Belgian province in West Flanders) with three cities in Great Britain, would be serviced by Sabena's propliners. From September 15, there would be eight weekly flights between Ostend and London. Then, from May 29, Sabena's propliners would connect Ostend with Manchester four times a week and three times a week with Glasgow. The Middle East routes would be flown daily with Sabena's Caravelle VIs. The French jetliners flew twice a week to Beirut via Vienna and Athens and twice to Istanbul and Tehran, with stopovers in Cologne and Athens. Cairo was provided with two flights per week via Vienna and Athens. There were also weekly flights to Tel Aviv via Vienna and Athens.

On the African network, the Belgium-Congo route, which Sabena operated in close cooperation with Air Congo, retained all flights from the previous timetable. It was six return flights per week with a Boeing 707-329 and a seventh weekly flight with Sabena's Douglas DC-7C until June 1. After June 1, only 707s were used on that route, and the weekly flight frequencies increased to eight. Six of those eight flights connected Brussels directly with Leopoldville, and the other two flights flew via Madrid, Geneva, Kano, Rome, or Athens. Two other flights were scheduled for the route to Elisabethville, which was: Brussels-Athens-Entebbe-Elisabethville.

There was hype about supersonic airliners in the early to late 1960s, like the Anglo-French Concorde, the American SST, and the Russian Tupolev Tu-144. Which airline would be the first to buy one, how many they would purchase, and what services they would offer with their speedy jetliners? On the other side of the coin, there were also noise pollution issues, the cost of fuel, the Concorde was a thirsty beast, and the cost of the tickets, among other things. Sabena was one airline that was not heavily involved in the supersonic dream. There were two main reasons for that. First,

Sabena was a small European airline with a modest route network compared to larger European carriers like Air France, BOAC, and Lufthansa and US mega carriers like United Airlines, American Airlines, Pan American Airways, TWA, Eastern Air Lines, and Delta Airlines. In the 1960s, the Belgian airline's flights were also less frequent. Most of its destinations were in Europe, with some routes to the Middle East and North Africa. Its long-haul routes were mainly two; the North Atlantic route focused on New York, some Montreal flights, and Mexico. The second was the African route network. There was the Johannesburg route, Burundi, and, of course, the Congo on that network. The other reason was that Sabena could not keep up with the frantic buying of jetliners in the new Jet Age. Sabena had to finance the purchase of its Boeing 707 and Caravelle jetliners entirely by loans, which were guaranteed by the Belgian State, of more than BFR 3.5 billion with an average duration of nine years and five months. Then, there was the issue of depreciation. The 707s and Caravelles would be almost redundant in the next ten years, but Sabena would still be paying off the loans for equipment that were past their prime. The purchase of Concordes would put the Belgian carrier in financial problems. Consequently. Sabena's management did not decide on the Concorde.

In June, Sabena released the first quarter results of 1964. In those first three months, Sabena transported 182,000 passengers, 18% higher than the first quarter of 1963. The ton-km used increased by 12%, and the ton-km offered by 5.5%. The load factor increased by 3.3%. The increases were 16% for passenger ton-km, 4% for freight ton-km, and 21% for mail ton-km. The first-quarter results of 1964 were 27% more advantageous than the previous year's results.

Air France and Sabena signed an agreement to operate in a pool for all their regular services between Brussels and Nice in July. Under the terms of the 5-year deal, Air France and Sabena would both share the services provided by the two companies, Caravelles.

Later in July, Sabena's general meeting of shareholders took place. The airline's turnover in 1963 amounted to BFR 4.2 billion, with a gross surplus of BFR 304.5 million. The relative improvement in the results of the 1963 fiscal year was primarily attributable to an increase in receipts that outpaced increases in expenditures and expenses. In 1962, total expenditures and expenses were BFR 4.48 billion, whereas, in 1962, they were BFR 4.23 billion. Operating revenue and miscellaneous receipts were BFR 4.2 billion in 1964, compared to BFR 3.76 billion in 1962. Compared to 1962, the expenses in 1963 climbed by 5.82%, production increased by 14.3%, operating revenue increased by 11.74%, and air traffic increased by 13.76%. The cost price per offered ton-km decreased by BFR 1.18, from BFR 14.02 in 1962 to BFR 12.84 in 1963. The IATA (International Air Transport Association) tariff cut from BFR 19.30 in 1962 to BFR 19.11 in 1963 resulted in a decline in operating performance earnings per ton-km. Overall, Sabena did much better financially in 1963 than in the economically damaging year of 1962.

In terms of air traffic, the Brussels-London pool agreement between BEA and Sabena was extended for another five years in November. The previous pool lasted for ten years. Sabena's flight frequencies were increased from 21 to 24 per week to match the capacity of BEA's 21 weekly Trident aircraft flights. Sabena's winter timetable for 1964 had only a few changes from the previous year's winter timetable. Interestingly, there was an increase in Belgian tourist traffic to Spain in the winter. As was the case during the summer, Sabena provided two weekly flights between Brussels and Malaga. The number of Brussels-Las Palmas flights via Lisbon was increased to two. In addition, the tri-weekly Brussels-Manchester service, for the first time in winter, was operated by Sabena's Caravelle VIs. The French twinjet replaced the Convair CV-440 Metropolitans on that route earlier. Its Caravelles operated Sabena's flights to the Middle East. The carrier provided two weekly flights to each of the following cities: Beirut, Istanbul, Cairo, and Tehran, and one flight per week to Tel Aviv, Israel. Further afield, in Africa, the Belgian carrier's

Douglas DC-7Cs continued to fly the Brussels-Bujumbura link via Cairo once a week. The 707-329s were scheduled to fly the Brussels-Elisabethville route via Entebbe, Uganda.

1965

In 1965, Sabena started the year with another milestone. The Belgian airline recorded the 7000th flight from Belgium to America since the opening of the transatlantic link in late 1964. Also, on December 29, 1964, the carrier received its 9th Sud Aviation Caravelle VI N. With the 6 707-329s

Photo 51. This is a rare photo of Sabena Caravelle VI 'OO-SRE' at Frankfurt, in experimental tail colors tested in 1972 but never implemented. The Caravelle was delivered to Sabena on March 24, 1961, and sold to the French carrier Catair in June 1976. (Photo by Michael Bernhard)

already in operation, which made 15 jetliners in service with Sabena in January 1965. Then in March, the Belgian company announced it had ordered 2 of the new model 707-300C mixed passenger-cargo versions. The first was delivered to Sabena on April 17. There was also one more Caravelle VI N due to be delivered to the carrier. One fact regarding Sabena's Boeing jetliners, the 10,000th Belgium-Congo flight, was made on March 14, 1965, by one of Sabena's Boeing 707-329s. Sabena's fleet in April 1965 consisted of the following aircraft: 6 Boeing 707-329, 1 brand new Boeing 707-329 C (the second 707-329C was delivered on March 23, 1966), 10 Caravelle VI N, 9 Convair CV-440, 2 DC-7 C (leased to Caledonian Airways), 2 DC-7F (converted into an all-cargo version by Sabena's technical services department), 3 DC-6 (including 2 in service with Air Congo), 6 DC-6 B (including one leased to Sobelair), 3 DC-4 (all 3 in service with Air Congo), 24 DC-3 (including six based in Tripoli, and six in service with Air Congo). Helicopters: 2 Sikorsky S-58 (purchased from New York Airways), 1 Alouette II. All up, a total of 69 aircraft.

Knowing that Sabena was a sucker for helicopters, the French aircraft manufacturer offered its new Super Frelon to Sabena for evaluation. The Belgian carrier said it was captivated by the flying attributes of the French Super Frelon helicopter, with its engine-out safety and its all-weather capabilities. After an IFR approach through an overcast sky with heavy cloud cover, the helicopter landed at the central heliport in Brussels. However, Sabena never bought the helicopter, as Sabena's helicopter services were suspended for good in 1967.

Sabena's first-quarter results for 1967 were mixed. Sabena's expenditure increased by 7%, and revenue by 6%. As a result, the projected deficit was estimated at BFR 27 million, increased by BFR 10 Million. Concerning air traffic, significant increases were recorded: on the Belgium-North American and Mexican route network, a rise of 13% was logged. When broken down, the Mexico City line had a sharp increase of 50%, Montreal's traffic increased by 27%, while New York was up by 7.8%. Flights to Germany went up by 24% in the European theater, Scandinavia posted a 21% rise, and Italy jumped by 50%. Still, the most significant increase in Western Europe was in the Netherlands, where the air traffic figure skyrocketed by 96%. Meanwhile, in Eastern Europe, flights were up by 25%. Europe recorded a decrease of around 15% in cargo traffic, while the freight sector grew by 30% on the North American network. In addition, Sabena's expenditure climbed by 61%, while revenue increased by 51% between 1960 and 1964. The increase in ton-km offered was 15%, while the overall load factor was at the break-even point of 62%.

The summer schedule for 1965, which took effect on April 1, saw some routes being strengthened and some general frequency increases for the summer season. On the European network, from May 1, the Yugoslav city of Dubrovnik on the Dalmatian coast was served once a week by Sabena's Caravelles. The route was only open during the summer and ran until September 30. Many more flights were introduced in 1965 compared to the summer of 1964. With Sabena's Caravelles, the UK now had three more weekly flights between Brussels and London. The number of weekly flights with BEA in the pool has increased from 42 to 45. An extra flight was added to the Manchester route, which was previously served by three flights. In Spain, frequencies on the Malaga line increased from 2 to 3, and Palma was provided with seven weekly flights in which Sabena and the Spanish airline Iberia operated in a pool. In the Canary Islands, Las Palmas received five weekly flights, two from Sabena and three from Iberia. Sabena's flights to Portugal, Spain's neighbor, were increased to seven per week. According to the Belgian carrier, all future flights would be operated in a pool with TAP, the Portuguese national airline. TAP would fly three of the trips while Sabena would fly four. In addition, Sabena flew to Morocco, one of the exotic destinations on its network. The north African country was served by three weekly flights in the summer of 1965, two of which stopped in Lisbon and the third of which flew via Madrid. Hungary was a popular destination in Eastern Europe; therefore, Sabena established a third weekly trip to Budapest, Hungary's capital. One of the flights involved a stopover in Prague. Four weekly flights from Brussels to Prague, Czechoslovakia's capital, were operated on Sabena's Caravelle jetliners. From May 24, Geneva was set to get daily Caravelle flights from Brussels. Geneva was an essential stopover for Sabena because it was one of the stopovers on the Belgium-Congo route, which meant that Sabena's Boeing 707 jetliners would fly there once a week.

On its North American route network, Sabena added an extra flight. Therefore, at the beginning of the summer season, the flights increased to 10. However, that number rose further on April 20, with 16 weekly flights offered. In Canada, the number of flights to Montreal doubled from 2 to 4, and there were two flights to Mexico City. On the Congo service, six weekly flights were made available by Sabena with the 707s in cooperation with Air Congo. That number increased to eight weekly flights from June 1. In addition, there were six weekly flights to Leopoldville and four per week to Elisabethville. The flights were either direct or with stopovers in Paris, Geneva, Rome, Madrid, Athens, Kano, or Entebbe. From May 1, the Air Congo service via Paris connected the French capital with both Leopoldville and Elisabethville. Concerning Johannesburg, the South African metropolis was still served by two Sabena flights each week. The weekly service between

Brussels and Bujumbura (Burundi) was inaugurated by Sabena and flew via Athens and Cairo with a Douglas DC-7C. In the Middle East, Sabena continued two weekly flights to each of the following cities, Tel Aviv, Beirut, Istanbul, and Tehran, with its Caravelles. In the land of the pharaohs, Cairo was the stopover on the Brussels-Bujumbura route. Two weekly Caravelle flights from Brussels were also added to the Egyptian capital.

Sabena released its air traffic performance results for 1964 in early February. The 14% rise in ton-km used from 203,039,266 in 1963 to 230,574,728 in 1964 indicated the improved traffic performance. The available ton-kilometers fell by 2%, from 392,890,628 to 385,812,943. Sabena carried more than 1 million passengers for the first time in 1964, 1,056,000 passengers to be exact, which was a 13% increase from 1963. Furthermore, cargo traffic continued to grow. Compared to the previous year, the cargo volume increased by 19% in 1964. It rose from 44,320,555 ton-km in 1963 to a total of 52,529,181 ton-km in 1964. New York held the World's Fair in 1964, which immensely increased Sabena's traffic on the North Atlantic network. In 1963, the sector recorded 71.58 million ton-km sold, but in 1964 it increased dramatically by 19% to reach 85.1 million ton-km. Moreover, the load factor of the North Atlantic route increased by 10.9% to 56.4%, a large jump compared to 1963. The number of passengers carried increased by 18%, and the seat load factor rose by 12.6%. However, due to the strong seasonal character of the North Atlantic route, Sabena could not reach the satisfactory utilization rate of 60%. In Europe, the cargo load factor increased by 4.6%, while ton-km used increased by 7%. On the African traffic, results were encouraging, especially concerning the air freight volume, which increased by 19% despite the turbulent political and ethnic situation in Central Africa.

In May, SAS and Sabena agreed to cooperate in providing nonstop flights between Brussels and Copenhagen. The two airlines were already in a three-way pool with KLM. However, their participation was permitted within the confines of the agreement. The nonstop flights were because Sabena was obligated to stop in Amsterdam on the way to Copenhagen with KLM in the picture. The new service was inaugurated on May 1 by an SAS Caravelle. The flights in the new pool were operated daily. With the new nonstop flights, neither Sabena nor SAS was obligated to make a stopover in Amsterdam, thus cutting the flight time between the Belgian and Danish capitals by one hour.

On May 26, 1965, Sabena reported that it had placed an order for two Concordes following the Concorde issue in 1964. The Belgian airline showed its determination to keep up with the rest of Europe's big carriers. But, the costs involved in the procurement and maintenance of such an expensive airliner would cause Sabena problems. The Belgian airline could not afford such luxuries. Moreover, its financial situation was not stable enough to operate an aircraft like the Concorde.

Photo 52. This is Fokker F-27-600 '00-SCA' in Sabena's classic livery, about to land at London Heathrow in September 1972. The aircraft was purchased by Sabena on August 12, 1969. On July 6, 1973, it was sold to Pakistan International Airlines. (Photo by Richard Vandervord)

Photo 53. A beautiful shot of a nose-down approach by this Sabena Fairchild FH-227B (US-built version of the F-27) at London Heathrow in March 1982. The aircraft was purchased by Delta Air Transport (DAT) and delivered to that airline on March 23, 1977. From 1979 to 1983, it was painted in Sabena's colors but operated by DAT on behalf of Sabena. In April 1983, Sabena 'returned' the FH-227 to DAT with no Sabena titles on the fuselage nor the curvy 'S' on the tail. (Photo by Richard Vandervord)

Sabena's financial results for 1964 saw the continuation of the financial reorganization of Sabena within the framework of the medium-term plan adopted in 1962. The amount of the intervention recoverable from the Belgian State was BFR 90 million lower than in 1963. In 1962 it was BFR 286 million. The receipts, for three years, increased more rapidly than the expenditure by 6.5%, and the traffic increased year over year by 13.4%. The average load factor rose from 51.7% in 1963 to 59.8% in 1964.

Sabena's winter timetable for the winter of 1955 to 1966 included a weekly Brussels-Tunis flight operated by its Caravelles. A Boeing 707-329C freighter in a mixed passenger/cargo configuration flew the route twice a week on the North Atlantic network. The 707-329C carried 90 passengers in economy class and five cargo containers. Sabena offered 14 weekly flights in the North American sector. In the previous winter, Sabena provided only ten flights. New York was the favored destination in 1965, with 11 flights a week, Montreal, 3 per week, and Mexico City, two flights in the African region, Sabena, and Air Congo provided eight weekly Europe to Africa flights, one more than the winter of 1964. The two airlines' flights connected six European countries to Central Africa (Belgium, France, Italy, Switzerland, Spain, and Greece). Two of Sabena's weekly flights to the Congo extended beyond Leopoldville to Johannesburg. In Europe, t Sabena's flight frequencies increased to the following countries: UK, Spain, Austria, and Switzerland. In the Sabena-BEA agreement, the two carriers' London-Brussels flights, operated in a pool, were increased to 40 per week, against 33 the previous year. In addition, there was an extra flight added to the Brussels-Manchester route, making it four per week. All of Sabena's flights were operated by Caravelles. Malaga was provided with four weekly flights by Sabena instead of 3 (in a pool with Iberia) in Spain. The Austrian capital Vienna, received an extra flight, making it six per week.

Sabena's cargo division was also busy in the winter of 1965. The capital of the Republic of Congo, Leopoldville, was included in the Belgian air freight network. Over the five winter months, eight Sabena airliners would operate cargo services to the Congo and the Central African region in general. According to Sabena's cargo timetable, 1,522 flights were planned for the coming winter period, of which 1,379 would be within Europe, 43 to Central Africa, and 150 to North America. The eight aircraft that had the task of providing Sabena's cargo operations in 1965 were the new 797-329Cs. In addition, the Douglas DC-6 all-cargo airliners combined would offer 16.97 million cargo ton-km, which was 54% more cargo capacity than in 1964.

1966

Sabena lost BFR 137.72 million in 1965, despite considerable growth in its operations. Mr. Deswarte, Sabena's Managing Director, stated that the dramatically increasing load on wages and social services resulted in the disappointing result. For example, between 1961 and 1965, Sabena's wages rose by an unacceptable 42%. The wage disputes were becoming more and more prominent as Sabena's aircrews took strike action on what seemed to be a regular basis, requesting more money from the carrier. Sabena was between a rock and a hard place financially. It could not entertain the desires of its staff for more money while the airline was making losses regularly. The airline had big plans for the future. It aimed to have an all-jet fleet of 30 aircraft, consisting of Boeing 707s and Caravelles, by 1970. However, with very tight monetary conditions and constant strikes hurting the airline, it would be challenging to pay for the new fleet and run air operations flawlessly with no hassles.

Nevertheless, Sabena's management was determined to have an all-jet fleet. Sabena had too many piston engine propliners in its fleet in the late 1960s. The DC-3s, DC-4s, DC-6s, DC-7Cs, and Convair CV-440 Metropolitans had to go. Turboprop and turbofan-powered aircraft were the way to go for the future.

One positive move for the Belgian carrier was the decision by its management to suspend more helicopter operations in 1966. Sabena discontinued its helicopter service to Maastricht on April 1. It also terminated the services to Liege, Cologne, and Duisburg. However, the flights to Rotterdam and Eindhoven were maintained. The cuts were three to four times higher than the traditional two-wing transport varieties because of the high costs of running the helicopter services. The decision to cease all helicopter operations in 1966 was at least ten years too late. If Sabena had ceased all helicopter operations in the first or second year of operations, when its management already knew that using helicopters to transport passengers was unprofitable, it would have saved itself a lot of money.

In February, Sabena placed an order for 4 Boeing 727-100 trijets. The first of Boeing's new medium-haul aircraft was scheduled to be delivered in March 1967, the second and third in April, and the fourth in May. Sabena chose two different versions of the Boeing 727-100. Two would be the standard all passenger version that could carry 102 to 118 passengers, and the other two were the 727-100QC version, where 'QC' stood for 'Quick Change.' The 727-100QCs were specially designed to transform from an all-passenger arrangement into an all-cargo configuration quickly. They had rollers on the floor and modularized passenger cabins to allow a much faster changeover time of 30 minutes. Once converted to an all-cargo version, the 727QCs could carry 18 tons of cargo. With the purchase of the 727s, Sabena's ground personnel, from engineers and aircraft mechanics, who had accumulated a lot of experience overhauling Sabena's Boeing 707-329s, would effortlessly maintain the new 727s.

Moreover, they already knew the Boeing way of conducting maintenance. If Sabena had bought the other trijet, the British 'Hawker Siddeley HS-121 Trident', the mechanics and engineers would have had to be trained from scratch. At the same time, part of the existing airport ground equipment for the 707s would be serviceable for the Boeing 727s as well. The 727-100s were powered by 3 Pratt & Whitney JT8D-7 engines. Each engine offered 14,500 lb. thrust. The 727-100s had a maximum cruising speed of 607 mph (977 kph) at 21,000 feet and 2,690 miles (4,330 km). Sabena planned on using the Boeing 727-100s for its routes to the Middle East and some destinations in Europe and Africa. The Export-Import Bank, the official export credit agency of the United States, granted Sabena a USD 17.5 million loan to purchase the 4 Boeing 727s.

Compared to 1964, Sabena's operating performance for 1965 showed a significant increase in traffic, which was noteworthy. Traffic performance improved by 12%, from 57.5% in 1964 to 58.3% in 1965. The total passenger numbers also went up in 1965, from 945,773 passengers in 1964 to 1,052,457, which was a jump of 11%. Each Sabena passenger flew 1,093 miles (1,759 km) on scheduled flights. Regarding cargo, Sabena's freight services division saw a significant increase in business in 1965. It had the best growth rate of the Belgian carrier in 10 years, at 24%. Also, in 1965, the amount of mail carried increased by 15%. The good news from 1965 seeped into the first three months of 1966. Sabena noted in its quarterly report that it achieved encouraging results in the first three months of the year. The number of passenger kilometers was 9% more than in the first quarter of 1965. While, the ton-km for cargo increased by 6%, and the transporting of postal items also increased by 7%.

In May, Sabena and Luxair inked a cooperation agreement under which the two firms agreed not to compete on the routes they jointly operated, particularly the Luxembourg-Brussels line, which would be conducted in a pool. As a result, the number of flights between the two capitals was dramatically increased, from seven to seventeen per week. In addition, the two airline companies stated that they would represent each other and provide airport services in their respective countries on behalf of the other company. The other victory for the Belgian carrier was to obtain the rights from the Greek Government to operate a weekly Sabena Caravelle service linking Brussels to Athens via Vienna and Thessaloniki for one year starting from April 1.

It finally happened. Sabena ceased all helicopter flights to the last bastion of Sabena's helicopter operations - the Netherlands, from November 1, 1966. As a result, the daily helicopter flights between Brussels, Rotterdam, Eindhoven, and Maastricht were terminated. Sabena's management decided to cease helicopter operations because investing in the renewal of the helicopter fleet was not economically justified. If Sabena had continued with its helicopter operations, it would have been necessary for the airline to replace its outdated S-58 helicopters with newer models, which would have been costly. That would render the operation of the helicopter services unprofitable for the airline. Consequently, Sabena sold all its Sikorsky S-58 helicopters to the Belgian army.

1967

Sabena's operating losses increased in 1966. According to its Managing Director, Mr. Deswarte, the losses were because of several factors. Two were more prominent. The first is Sabena's meager occupancy rate. Mr. Deswarte blamed the low load factor on Sabena's airliners on the Belgian business people and the general public for not using their national airline to travel. The second factor was the unwelcome development of relations with the Congo, where the Government of the African country blocked Sabena from withdrawing its considerable capital from the Congolese bank. It caused Sabena's management a lot of anguish. First, the Belgian carrier stated it suffered a loss of BFR 344.4 million due to the 1963 devaluation of the Congolese franc, representing 46% of the Belgian airline's capital. Then came the second blow. Sabena experienced severe difficulties transferring its revenue earnings from the Congo to Belgium in 1966. In 1965 it could move 67% of its revenue. But, in 1966, the percentage of funds allowed to be transferred fell even further to 40%. By the end of 1966, Sabena had a total of BFR 1.1 billion in a blocked account in the capital Kinshasa, which the airline wanted to claim; however, the Congolese Government refused to transfer the funds to Belgium.

Sabena's operational statistics for 1966 showed an increase of 7% in traffic performance over the previous year. The ton-km sold increased from 255.7 million in 1965 to 273.3 million in 1966. At the same time, the amount of ton-km offered increased by 8%, from 424.2 million to 458.4 million. On the other hand, the load factor declined slightly by 0.7%, from 60.3% in 1965 to 59.6%. The number of passengers transported in 1966 rose from 1,144,631 in 1965 to 1,175,661, a growth rate of 3%. Also, Sabena's cargo division did well in 1966. In 1965 Sabena sold a total of 60.98 million cargo ton-km, and in 1966 the figure increased to 71.98 million, an increase of 18%. Mail volume also went up by 3%, from just over 6 million ton-km in 1965 to 6.17 million in 1966. Mr. Deswarte stated that Sabena's air traffic in the European sector was satisfactory, with a 7% increase in the ton-km sold and an 8% rise in ton-km offered. Two other areas, Mexico and South Africa recorded the highest traffic growth. On those two routes, the number of ton-km sold increased by 10%, while the number of airfreight ton-km sold increased by 35%. Additionally, the number of passengers carried went up sharply by 11%. Regarding the North American sector, the results for 1966 were practically the same as 1965, with one exception, the volume of air freight in that sector rose by 6%. In January, Sabena celebrated another record by its legendary Caravelles. After just six years in service with Sabena, the Caravelles flew 2 million passengers and acquired 100,000 flight hours. They also made 123,600 take-offs and landings in the first six years, an average of 60 take-offs and landings per day. Even more impressive was that during the 1966 summer season, Sabena's 10 Caravelles accounted for an unbelievable 82% of the carrier's air traffic on the European and Middle Eastern networks. The Caravelles were the gifts for Sabena that kept on giving.

Photo 54. Sabena Boeing 727-29C 'OO-STB' with classic colors, on short final at Frankfurt Airport on April 14, 1973. Delivered to Sabena on June 6, 1967, and sold to the Belgian Air Force in January 1976. (Photo by Michael Bernhard)

Photo 55. Another shot of 'OO-STB' with its last paint job, taxiing at Vienna International Airport on April 20, 1974. (Photo by Michael Bernhard)

In its 1967 summer timetable, Sabena had more flights to offer. The number of flights to Tel Aviv was doubled, and flights to Montreal, where the World's Fair was held, increased. The number of weekly flights between Brussels and New York also rose. From June 5 to September 27, there were 14 to 16 flights per week on that route. Some of the flights from Brussels departed in the morning to allow passengers to make the appropriate connections for onward flights in the United States on the same day. In 1967, Sabena provided nonstop flights to Stockholm and Copenhagen, which pleased its passengers. Previously some of Sabena's flights stopped over in Amsterdam, which meant that Sabena's passengers flying to the two Scandinavian cities had to endure an hour on the ground at Schiphol Airport. One European city provided with more flights was the Hungarian capital Budapest. Meanwhile, there was some positive news on Sabena's services to the Netherlands in 1967.

After Sabena shut down its helicopter operations to Holland in 1966, the Belgian carrier stated it would offer flights to Belgium's northern neighbor with a fleet of small aircraft on the discontinued helicopter routes. In addition, Sabena said it would provide flights from Brussels to Rotterdam, Eindhoven, and Maastricht from the summer of 1967. Sabena hoped to get all the Dutch passengers back and fly with the Belgian carrier with its small aircraft, as they had done with its helicopter services. During the first quarter of 1967, Sabena reported that its operating results showed a significant improvement in the first quarter of 1967. During the first three months of the year, the number of ton-km offered increased by 29% compared to the previous year, and the number of ton-km used went up by 41%, which was 4.3% higher than in 1966. In addition, Sabena saw more passengers use its services in the first quarter of 1967. Compared to the previous year, the number of passengers carried on its regular scheduled services increased by 16%, and its load factor rose by 6.7%.

There was a rumor in aviation circles in mid-1967 that Sabena and KLM were about to sign an agreement whose aim was to unite the two airlines. In actuality, it was not a rumor but a fact. The two airlines sought to get closer, which made sense. They were part of a political-economic alliance, the Benelux Union, a formal international intergovernmental cooperation of 3 neighboring countries, Belgium, Netherlands, and Luxembourg. However, the airlines of the three nations had little in common. KLM was the giant of the three. Its immense worldwide route network and sheer size of its fleet cast a large shadow over Sabena and Luxair. Sabena's financial situation was particularly severe in 1966, and it was difficult for the Belgian company to see a comeback in the years ahead. Sabena had only two options: either cut back on its loss-making activities and cut itself off from its least profitable routes or partner with a financially sound airline like KLM. Otherwise, it could join a group of airlines to integrate into a larger whole where cost-cutting measures would be easier to carry out and more profitable.

Due to Sabena's difficult financial situation, a reorganization was drawn up by Sabena's management and implemented, which improved the problem to a certain extent. But, the bottom line was money; Sabena did not have any. So, what did the management do in that case? It culled some of its personnel and performed a rationalization of Sabena's airline fleet. Meanwhile, KLM was in an ideal financial situation. In 1967, the airline was prepared for expansion and modernized and enlarged its fleet. It had ordered several Douglas DC-9s, 4 DC-8-63s, and 3 Boeing 747s. Most importantly, KLM could make all those aircraft purchases with its funds, which was far from the case with Sabena. Thus, any agreement signed between the two airline companies, even though their governments wanted it, was a pipedream, at least in the year 1967 it was. Had a thorough check of Sabena's inventory taken place before negotiations began, it would have proven too unfavorable to Sabena. Moreover, Sabena was in financial and governmental disputes with its erstwhile colony, the Congo, which would not end anytime soon. In the end, the talks ended with no results.

In July, Sabena released its financial results for 1966. In that year, the carrier's total turnover decreased by 1.1% compared to 1965 and reached BFR 4.96 billion. Meanwhile, its revenue surplus amounted to BFR 136.2 million. However, after deducting the necessary amortizations of BFR 647.7 million and the statutory dividends to be paid on the share capital of BFR 37.5 million, the net deficit amounted to BFR 576 million, by which the government had to foot the bill once again. In addition, Sabena was under pressure from 1966 to 1967 due to the Congo situation and strikes by most of its flight crews, who demanded higher wages and social security contributions. On April 28, at Brussels Airport, the first of 4 Boeing 727 trijets that Sabena had ordered arrived from Seattle. Sabena's new Boeing 727-29C was a mixed version type with 102 seats, 90 in economy class and 12 in first class.

Sabena put its Boeing 727s in service on June 26, between Brussels and Milan and between Brussels and London. The jetliners provided daily flights on each of those two routes for the remainder of the summer season. Then, on July 31, the Boeing trijet entered service on the Middle East route network. Sabena offered two weekly flights to Istanbul and Tehran via Vienna and Athens and two to Beirut via Vienna and Athens. The other route was to Cairo via Cologne and Athens, and a weekly flight to Tel Aviv was made available. Finally, at the beginning of September, Sabena provided two weekly services; the first was the Brussels to Jeddah line, the first time Saudi Arabia had been offered a regular service from Brussels. The second service was the Brussels to Bujumbura route via Athens. Indeed, that must have been the longest route that any 727-100 had to fly. In addition, Sabena's new 727-29QC 'Quick Change' airliners were put into operation on Sabena's cargo network on the Brussels-London route. Also, the 727s routes were expanded to the following European cities, Manchester, Stuttgart, Frankfurt, and Paris, in the ensuing weeks.

In August, Sabena quickly put in new orders for Boeing jetliners. It ordered one Boeing 707-320C and a third Boeing 727-29QC, with delivery scheduled for October. Sabena's fleet at the beginning of September 1967 was as follows: 6 Boeing 707-329, 3 707-329C, 2 727-29C, 2 727-29QC, 10 Caravelle, 9 Convair CV-440, 2 DC-7CF, 3 DC-6B, 8 DC-3. Light aircraft: 9 Cessna 310, Helicopters: 2 Sikorsky S58C, one Alouette 2.

On the topic of airliners, one of Sabena's 707s created history on August 30, when it set a new flight record on the Brussels-Johannesburg route, 10 hours 31 minutes without a stopover, a total distance of 5,516 miles (8,877 km). Even with the modern airliners today, the usual time duration on that route is 12 hours 55 minutes; now that was fast!

On July 18, 1967, Sabena held its Annual General Meeting. Sabena's turnover in 1966 was down 1.1% from 1965, at BFR 4.96 billion. Nevertheless, the company made a gross profit of BFR 136.9 million after depreciation of BFR 674.7 million, and the statutory capital remuneration was BFR 37.5 million. As a result, the State absorbed a net loss of BFR 576 million. The 1966 fiscal year saw Sabena's financial situation deteriorate sharply, even though Sabena's traffic results had been improving continually since 1962. The reason for the carrier's losses was plentiful. First, there were still the difficulties Sabena was experiencing with Air Congo concerning technical cooperation issues and the blockage of its funds by the Congolese bank. Then there was the strike of the majority of the aircrew and its consequences, which transpired from May 20 to June 25 and resulted in a significant increase in labor costs, and social security contributions. All those events contributed to the financial distress that Sabena faced in 1967.

In October, Sabena announced its winter schedule. For the winter season, Sabena added two additional destinations: Bombay, India, and Nairobi, Kenya, to its network, bringing the total number of cities covered by Sabena to 59. Of which five were in Asia and twelve in Africa. A Sabena 727 operated the weekly route from Brussels to Bombay. With stops in Vienna, Athens, and Tehran, the flight took 10 hours and 30 minutes. At the same time, Nairobi was given a weekly flight that took nine

hours and included a stopover in Vienna. The 707s served Nairobi, which was a stopover on route to Johannesburg. The flight was one of two weekly to Johannesburg. The second flight went to South Africa via Madrid and Luanda, Angola.
Sabena announced in October that its flights to Kigali, Rwanda, and Bujumbura, Burundi, would be operated exclusively with the 707 or 727 jetliners. Sabena's destinations in Africa in October 1967 were as follows: Casablanca, Tunis, Kano, Lagos, Abidjan, Kinshasa (Sabena served Kinshasa with three weekly flights and two by Air Congo), Bujumbura, Kigali, Nairobi, Entebbe, Dares-Salam, and Johannesburg. In Europe and the Middle East, the two Dutch cities of Eindhoven and Rotterdam were included once again in Sabena's route network on August 28. In September, the Belgian carrier inaugurated a new service to Jeddah, Saudi Arabia. Las Palmas was to be serviced by Sabena's 727s,

Photo 56. A nice shot of Sabena's Caravelle 'OO-SRI' on short final at Heathrow Airport, London, in May 1971. This Caravelle was the ninth delivered to Sabena on December 29, 1964. Its last year of operations was in 1978, with Sobelair. (Photo by Richard Vandervord)

making the Canary island destination the 18th one for Sabena's trijets. On the Brussels-Nice route, there was an increase of flights on February 1, 1968, to 8 per week. That was due to the yearly winter Carnival period. Air France and Sabena operated the route in a pool.

1968

Sabena's air traffic performance results for 1967 came out in January. They showed that the airline reduced its deficits considerably to BFR 193 million, which was about a third of the debt recorded in 1966. Sabena's Managing Director, Mr. Deswarte, stated that the improvement resulted from a significant increase in air traffic, by 18%, and some internal measures, such as drastic budget cuts, improved sales, and a cessation of personnel recruitment. Some external factors worked against the Belgian carrier; however, the Six-Day War in the Middle East, the volatile political events in Greece and Nigeria, and the constant difficulties in the Congo. Without those setbacks, the company would have reduced the deficit for 1967 to less than BFR 100 million and probably even closed the 1967 financial year in the black. Alas, it was Sabena's fate to suffer another loss in 1967.

Regarding the high increase of Sabena's air traffic in 1967, the main improvement came from the North Atlantic sector, where traffic increased by 33%, and the load factor went up by 1.8%. However, the most impressive rise was from Sabena's cargo division which posted a 49% increase in its cargo services between Belgium and Africa. The number of cargo ton-km sold increased from 273.3 million in 1966 to 322.6 million in 1967, while the number of cargo ton-km offered also rose by 18%, from 458.4 million to 542.7 million. In passenger numbers, Sabena carried 1,243,146 passengers in 1967 compared to 1,175,661 in the previous year, an increase of 6%. The average distance traveled by each passenger in 1967 was 1,161 miles (1,869 km). However, it was the cargo numbers that had Sabena's management salivating. The number of freight ton-km sold in 1967 was just under 106 million ton-km (105,976,625 to be exact) compared to 71,981,088 in 1966. Postal freight also increased by 8%, from 6.2 million ton-km to 6.7 million. The cargo load factor on the North Atlantic route also improved by 1.8%. Sabena's cargo network was busy in the forthcoming summer period. The carrier established a timetable for its cargo services, flown by six aircraft, including 3,238 flights in Europe and 183 to Africa. There were a further 245 flights to the North Atlantic over the summer. Finally, in 1968, Sabena's impressive 49% rise in air cargo services to Africa was the talk of the town. In April, Sabena announced that it had put in an order for three more Boeing 707-329Cs. Boeing said it would deliver the airliners from mid to late 1969. Sabena's order brought to 19 the number of Boeing jets in service with Sabena. The Belgian carrier also placed an option with Boeing for two 727-29s, but it decided not to proceed with the order. The Sabena fleet in April 1968 was as follows: 6 Boeing 707-329, 3 707-329C, 10 Caravelle 6R, 1 DC-7CF, 3 DC-6B, 8 DC-3, 9 Cessna 310, 2 727-29, 2 727-29QC. On order: 1 727-29QC, 1 707-329C, and an option on two Concordes.

In July, the Annual Ordinary General Meeting of Sabena Shareholders took place. The results for the 1967 fiscal year demonstrated a significant improvement over those of 1966 because of the aggressive efforts by Sabena's management to reestablish the carrier's financial stability. Despite a decrease in unit income, overall revenue increased by 14%. Expenses increased by less than 6%, reflecting Sabena's fairly spectacular recovery in 1967, which would contribute to the company's future advancement. The revenue figure of 14%, in particular, reflected the company's well-executed approach, allowing it to prepare for an increase in unit revenue. Sabena's turnover in 1967 was BFR 6.35 billion, or BFR 695.3 million more than it was in 1966 when it was BFR 5.65 billion.

During the winter season, Sabena and Air France shared flights on the Brussels-Paris route. Sabena would fly three times a day from Monday to Friday and twice on Saturdays and Sundays. Meanwhile, the French national airline would operate two additional flights. In late November, Sabena announced the launch of two new routes to Tokyo: one via the polar route, with a stopover in Anchorage, and the other via Southern Asia, with stops in the Middle East, Bangkok, and Manila. Sabena's 707s would fly both routes. In addition, there was some movement in Sabena's cargo routes. From November 1, 1968, five new stopovers were added to the already important cargo network: Montreal, Rome, Alicante, Kigali, and Basel.

The Austrian carrier Austrian Airlines (AUA) and Sabena signed an agreement to jointly operate transatlantic services, starting from April 1, 1969. Thanks to the mutual understanding, Austrian Airlines provided a daily flight from Vienna to New York, with a stopover in Brussels. The aircraft was a leased Sabena Boeing 707-329 that bore the Sabena and the Austrian Airlines logos. Austrian Airlines provided half of the seats, 73 in total, divided into first and economy class, as well as half of the cargo space for the joint air service. The same went for Sabena. The services were provided in the cabin simultaneously by the two airlines' hostesses, who were dressed in their respective uniforms. The two airline companies complemented each other with the new combined service because it allowed AUA to fly on the lucrative North Atlantic route. At the same time, Sabena could provide its passengers with interesting connections to Eastern European countries because Austrian Airlines had the most extensive network in that sector.

The death of Sabena's long-serving chairman of the board, Gilbert Perier, on March 13, 1968, was a sad moment for the company. Mr. Perier joined Sabena as secretary of the board of directors in 1930. He was appointed secretary-general in 1932, just two years after joining the company, and administrator in 1936. During WWII, he fled to the Congo. He held two positions: one as Sabena's managing director and the other as director of the 'Congo Company of Commerce and Industry' Leopoldville headquarters, where he had been its representative since 1939. In 1943, he departed the Congo for London at the request of the Belgian authorities. He was appointed Chairman of Sabena's Board of Directors in 1947. In addition, from 1948 to 1949, he served as Chairman of the International Air Transport Association (IATA). In July, Jean Van Houtte, Belgium's former Prime Minister and Finance Minister, took over as head of Sabena.

1969

The new head of Sabena did not mince his words. In March, Mr. Van Houtte stated that because of the company's focus on the Congo and the subsequent events there, Sabena had to go through a rough patch, but that was to be expected. Sabena suffered the most severe damage of any Belgian company due to the Congo's independence. Nevertheless, it had already made a significant recovery, with a 25% reduction in the workforce and a 25% increase in output. Mr. Van Houtte stated his support for the formation of Air Union as early as 1948. Mr. Van Houtte desired a European airline alliance based on the realities of each participating airline's resources and level of participation in European policies. Mr. Van Houtte did not eliminate competition among European airlines; however, he stated that, from a European standpoint, competition among European companies had gone too far, stifling their ability to compete with the American airline giants. As a result, no European carrier considered a deep, long-term alliance in the late 1960s.

According to Sabena's 1968 air traffic results, it carried over 1.3 million passengers for the year. In addition, there was a 22% increase in the freight sector over the previous year. Meanwhile, Niamey, Tangier, Guatemala, Alicante, and Cairo had been added to Sabena's route network, bringing the total number of destinations on the airline's route network to 64. Sabena had 24 jets in its fleet in 1969, with the largest, the incredible 747 Jumbo Jet, set to join the fleet in 1970. Sabena placed an expected but long-delayed order for two Boeing 747s for delivery in November and December 1970 in February. The new European 747 Club of Air France, Alitalia, Lufthansa, and Sabena signed an agreement to maintain their Boeing 747 fleets on March 14 jointly. All four airlines would be able to use a flight simulator installed by Lufthansa in Frankfurt.

Sabena began a weekly 5:20 p.m. flight from Brussels to Tokyo, stopping in Anchorage en route, just in time for the summer season, in the first week of April. A second service to Japan via South Asia would begin operations later in August. In addition, thanks to the Sabena-Swissair pool, there were four daily flights between Brussels and North America and six flights to Switzerland, three to Geneva, and three to Zurich. In 1969, the Belgian company operated a 54,810-mile (88,207-kilometer) freight network connecting Brussels to 31 cities, including 17 European cities, seven African cities, three North American cities, and 4 Asian cities. Sabena's cargo planes included five 707-329Cs, three B-727 QCs, one C-47, and one Fokker F-27. The cargo division offered 92 million ton-km or more than 22% of the airline's total production. Sabena planned to transport 22% more goods than in 1968, or 35,000 ton-kilometers, during the 7-month summer season of 1969. As a result of the numerous additions, the Sabena network spanned 101,283 miles (163,000 km) and 65 cities in April 1969. At the same time, Sabena's fleet consisted of the following aircraft: 6 Boeing 707-329, 3 707- 329C, 2 727-29, 3 727-29QC, 10 Caravelle 6, 2 DC-7CF, 3 DC-6B, 7 DC-3, 9 Cessna 310. It had five aircraft on order; 3 707-329C and 2 747 Jumbo Jets. In addition, it had an option for 2 Concordes.

The Belgian national carrier had to close the 1968 financial year with a loss of 227 million. In 1967, the loss was BFR 190.2 million. Nevertheless, the turnover reached BFR 5.7 billion, 1% more than in 1967. As a result, the gross operating surplus was BFR 275.6 million, but after a depreciation of BFR 552.7 million, the result was a BFR 277 million deficit. On the other hand, Air traffic increased by 7.7% in 1968, with 0.5% for passengers and 22.9% for goods, but postal traffic decreased by 6.5%. According to the annual report, the reasons for the downturn were rising production costs and a decrease in Sabena's share of North Atlantic traffic due to stiff competition.

Photo 57. Sabena 707-329C 'OO-SJN' at München-Riem (Munich) Airport on August 3, 1977. Delivered to Sabena on July 22, 1969. It was sold to NATO on July 4, 1988, and converted to a CT-49A. (Photo by Stefan Röhrich)

The Future Awaits
1970-1979

1970

It was the start of another decade, and Sabena strived to improve. It did so quite emphatically. Its air traffic in 1969 increased by 17% over 1968. The number of passengers transported by Sabena the previous year reached 1,374,952, compared to 1,257,519 in 1968, representing a jump of 9%. Freight transport continued to grow at a breakneck pace in 1969 with a 36% increase, compared to 1968. Sabena carried 176.8 million ton-km of freight in 1969, up from 130.3 million in 1968. It should be noted that freight transport accounted for more than a third of total ton-kilometers transported by the company in 1969, having almost tripled in the previous five years. Production increased by 17% from 581.1 million ton-km in 1968 to 682.1 million ton-km in 1969. However, the airline's load factor remained virtually unchanged, falling from 59.8% in 1968 to 59.6% in 1969. Sabena struggled financially in 1969. Once again, the Belgian national airline lost BFR 485 million, and the deficit was taken care of by the state, which covered Sabena's total losses to the tune of BFR 6.7 billion.

The Belgian airline's network expanded with several new stopovers in Asia, Africa, and Belgium in 1969. The most spectacular development was opening new routes to the Far East. Sabena provided two weekly flights to Tokyo. The first flight was inaugurated on April 4, 1969, and flew to Japan via the 'Polar' route over the North Pole and included a stopover in Anchorage, Alaska. The second line was the Belgium-Japan link that opened on August 3, 1969, and flew via southern Asia to Japan. In April 1970, Sabena opened a third Brussels-Tokyo route which was as follows: Brussels-Athens-Tehran-Bangkok-Tokyo. Thus, Tokyo would be reached by the south Asian route twice per week and once per week via the 'Polar' route. The total flight distance of the 'Polar' and South Asian routes was 18,020 miles (29,000 km). Therefore, it could be considered a round-the-world route.

As for flights to North America, 3 to 4 daily flights, or 23 per week, were provided in the summer between Brussels and New York. For Canada, there were five weekly flights to Montreal. Further south in Central America, two weekly flights from Brussels were scheduled to Guatemala and Mexico City every Tuesday and Saturday.

In the south Asian region, Beirut and Bombay were offered one flight per week, while Jeddah, Tehran, Tel Aviv, and Bangkok all had two flights per week. Tokyo was the lucky one with three weekly flights. As for the African continent, a fifth weekly flight became available between Brussels and Kinshasa. The first two flights were direct, while three stopped over in Geneva and Athens. As a result, Sabena was fast becoming a popular airline in Africa. In addition to its two flights to Johannesburg, the Belgian carrier also provided flights to the following African cities: Entebbe (Uganda), Nairobi (Kenya), Lusaka (Zambia), Bujumbura (Burundi), Dar-Es-Salaam (Tanzania), and Kigali (Rwanda), all in eastern Africa. As for the western part of Africa, Sabena also provided flights to Kano and Lagos (both in Nigeria), Niamey (Niger), Abidjan (the Ivory Coast), Conakry (Guinea), and Monrovia (Liberia). On top of that, Sabena offered the following destinations in Northern Africa: Cairo, Casablanca, Tangier (both in Morocco), and Tunis.

In March 1970, Sabena and SAS agreed to the joint operation of flights between Belgium and Scandinavia to offer their customers a service better suited to their needs. The agreement, which commenced on April 1, 1970, resulted in increased cooperation between the two companies on the commercial level and allowed them to coordinate the timetables of their Brussels-Copenhagen-Stockholm routes better. As a result, from the start of the summer schedule, Sabena and SAS offered two daily flights between Brussels, Copenhagen, and Stockholm, one of which was in the morning and the other in the afternoon. In addition, SAS provided the morning service on the following route: Stockholm-Copenhagen-Brussels, while Sabena operated the afternoon flight between Brussels and Copenhagen. It was convenient to travel to Stockholm or other cities throughout Scandinavia from Copenhagen. The agreement also stated that from April 1, 1971, the two airlines would operate three daily services (morning, noon, and evening) jointly.

Sabena's 1970/71 winter timetable was the first to introduce the entry into service of the Boeing 747 on the North Atlantic route network. Sabena's 747s would fly daily from Brussels to New York, while one of Sabena's 707-329 would operate a second daily flight. Although the 'Jumbo' was not commissioned until two months after the start of the winter period, the Boeing 747s accounted for a large chunk of the transport services on the North Atlantic route. They provided almost 50,000 of the 112,000 seats on Sabena's aircraft. The Brussel-New York flight with the 747 only took seven and a half hours, 30 minutes faster than the 707s. Sabena was scheduled to fly 14 flights per week to New York, seven with the 747s and seven with the 707s. There was no change on the Montreal route, as that city had the same number of flights as in the summer timetable - five a week. The same went for Mexico and Guatemala. Sabena offered them two weekly flights each.

With the inclusion of Manila in Sabena's East Asian route network, a direct air link was established for the first time between Belgium and the Philippines. All three routes that Sabena flew to Asia all ended in Tokyo. Meanwhile, In Africa, Monrovia, the capital of Liberia, was provided with two weekly flights, one more than the previous timetable. Previously 727s were flying that route, but Sabena replaced the medium-haul jetliners with its 707s. As a result, during the five months of the winter season, Sabena offered just over 900,000 seats on its scheduled aircraft, which performed 12,500 flights, representing an 11% increase in ton-km compared to the previous year.

In November, in Sabena's cargo division, the carrier announced that, in response to the ever-increasing expansion of air freight transport, it modernized its freight handling system at Brussels International Airport at the cost of BFR 2.5 million. The importance of cargo services at Brussels airport, which totaled 79,033 tons in 1969, necessitated modernization. Accordingly, the airline implemented two distinct measures for a more efficient process. One was to double the storage capacity of small packages, and the other to make large consignments easier to handle. Sabena was also investigating the construction of a new terminal, which would initially accommodate 350,000 tons of freight per year, compared to 120,000 tons in the current cargo terminal. The larger terminal was scheduled to open in two years.

With the new 747s ready to arrive in December from Everett, Washington state, and with more jetliners at its disposal, the Belgian airline was determined to expand its long-haul network further by opening new routes to South America. Earlier, the Belgium government had signed an air transport agreement with Senegal, Argentina, and Chile, and Sabena's route planners were thinking of a Brussels-Dakar-Buenos Aires-Santiago weekly service from April 1, 1971. With the tourists, business people, and exporters in mind, Sabena figured it would be an excellent idea to open the route, which the Boeing 707s would fly. The creation of a direct air link between Belgium and South America was justified by the growing volume of exchanges between Europe and the countries of Argentina and Chile. Moreover, the two South American nations were increasingly turning towards Europe for economic reasons and, in particular, the European Economic Community.

Sabena officially introduced its first jetliner, the Boeing 707-329 Intercontinental, to its long-haul network ten years earlier, on January 19, 1960. It was the first airline to provide flights from Europe to Africa on its new shiny jets. It was also the first European airline to operate a jet service to the US when it introduced the Boeing 707 on its Brussels-New York route a few days later. Just over ten years later, on November 26, 1970, Sabena took delivery of its third Boeing jetliner, after the 707 and 727. It was not just a run-of-the-mill airliner; it was the 'Queen of the Skies', the inspirational 747. The 'Jumbo' did not start operations until January 8, 1971, when it took off for its first New York flight. Sabena provided daily747 flights between Brussels and New York from that day.

Interestingly, Sabena introduced three seasonal seating arrangements for its Boeing 747s on the New York route. The first configuration, a temporary one, was in place from January 8 to June 14. The 747 was outfitted to accommodate 245 passengers in economy class and 32 in first class. Furthermore, an area at the back of the cabin was partitioned off to accept containerized passenger baggage. That allowed the entire underfloor volume to be used for cargo. In this arrangement, the 747 would carry 20 tons of cargo. Things would change for the high season in summer. In the second configuration, used from June 15 to September 14, the Jumbo was fitted with 365 seats in economy, with no change in the first-class seating and no baggage containers in the cabin. At the end of the summer season, the last configuration came into play from September 15 to October 31. During that period, the 747 was fitted with 277 economy class seats, with no changes for first-class, as before.

Before introducing the 747 to the fleet in August, Sabena announced a new order with McDonnell Douglas. The Belgian carrier put in an order for 2 DC-10-30s. It was the first time that Sabena had ordered a Douglas airliner since the DC-7C in 1955. The Sabena DC-10-30 was equipped with 3 General Electric engines and offered accommodation for 250 passengers. Delivery of the 2 DC-10s was to take place in 1975. This was Sabena's fleet in April 1970: 2 747-129, 6 707-329, 6 707-329C, 2 727-29, 3 727-29QC, 10 Caravelle, 2 DC-6B. Light aircraft: 9 Cessna 310, 6 SIAI SF260. It also had the 2 Concordes on order.

The ATLAS Group

An essential conference with Europe's ten leading airlines took place in June 1967 at Air France's headquarters at Maine-Montparnasse, in Paris. The airlines at the meeting were: Swissair, Air France, Aer Lingus, Alitalia, BOAC, Lufthansa, Iberia, KLM, SAS, and Sabena. They created the 'Montparnasse Committee,' a coalition whose goal was to find ways to foster tighter cooperation between the airlines to achieve the highest levels of efficiency and profitability. They inked a technical cooperation agreement. The carriers agreed to coordinate their long-term equipment policies and cooperate in the operation of big subsonic and supersonic aircraft, like the Boeing 747 and Concorde.

Following another meeting held in August 1967 on the possible purchase of Boeing 747s, three of the members of the 'Montparnasse Committee', Swissair, SAS, and KLM, took their own path and formed a group known as the 'KSS' Consortium ('KSS' stood for KLM-Swissair-SAS). The KSS group held a second summit in May 1968. All three carriers had placed firm orders for the 747s. As a result, the three airlines held a meeting to determine how they would collaborate on the overhaul tasks for their Boeing 747s, such as airframes, engines, and other critical equipment. They agreed to divide the workload evenly while reducing their financial obligations. Four of the airlines in the Montparnasse Committee met in early 1968 to exchange views. They were Air France, Alitalia, Lufthansa, and Sabena. Air France launched the idea of cooperating in the operation of future aircraft procurement following a prospective analysis of the situation in the forthcoming ten years. Air France found that, in the future, planes would be larger than current planes, and their transport capacity would be at least double that of the airliners in 1970. The French carrier said that regardless of its size and the expected increase in its air traffic, even a large company will only own a small number of the large jetliners. Therefore, the best way to deal with the purchasing and overhauling of the aircraft was to work together. The five 'commandments' for the Atlas Group were as follows: 1) Only purchase airliners that were identical or close enough. 2) All airlines in the group would have to share workloads fairly. 3) There should be a fair distribution of costs. 4) Any airline could participate in one or more of the activities. And 5) The ATLAS Group must have mutual trust in each other.

Photo 58. Sabena's second 747, 'OO-SGB,' is seen here, in classic colors, at Munich International Airport in October 1974. Delivered on December 18, 1970. It was sold to Nationair on December 14, 1990. Nationair then leased it to Garuda of Indonesia. (Photo by Norbert Kröpfl)

Photo 59. Sabena 747-129 'OO-SGB' this time in the new Sabena1980s scheme, at Chicago O'Hare International Airport on June 14, 1986. (Photo by Richard Vandervord)

Photo 60. Another image of 'OO-SGB' taxiing at John F. Kennedy International Airport in New York in August 1975.
(Photo by Richard Vandervord)

Photo 61. 'OO-SGB' taking off on July 16, 1981. (Photo by Udo K. Haafke)

Photo 62. Sabena's Jumbo on final approach to London Gatwick in September 1986. (Photo by Richard Vandervord)

Photo 63. 'OO-SGB' at London Gatwick in May 1987. (Photo by Richard Vandervord)

Consequently, a memorandum of understanding was signed on March 14, 1969, by the Atlas Group. Atlas was formed to enable the introduction of future subsonic and supersonic aircraft into the respective fleets of the Group and their use in optimal technical and economic conditions. First and foremost, the joint efforts focused on the overhauling and maintaining of the Boeing 747s, which all the airlines in the group had purchased, as well as the operational activities associated with the Jumbo, which included, among other things, crew qualification, training on a joint 747 flight simulator. Thus, each partner's tasks were as follows: Air France was in charge of the airframes, Lufthansa was assigned to overhaul the engines, and Alitalia and Sabena for some critical components. All the airlines in the group assumed their tasks very effectively. Motivated by this positive outcome, the Atlas Group decided to assess other future aircraft in common. The first two aircraft that the group wanted to purchase were the large long-haul trijets available on the global market at the time, namely the McDonnell Douglas DC-10 and the Lockheed Tristar. The ATLAS Group chose the DC-10 after a lengthy process. Specialized groups were immediately put to work defining the Atlas specifications for the manufacturer. They explained to them what the carriers in the group required and how the manufacturer could produce an aircraft that was more in line with the needs of the Atlas Group's airlines. Similar studies were launched and submitted to the European Airbus manufacturer to define what type of airliner the group had in mind. The group also looked to expand its technical cooperation into other fields and with new partners. The ATLAS Group t was a determined group and one that Sabena was glad to join.

1971

After 48 years of international services, Sabena decided to open a domestic route network. From April 1, 1971, Belgians could finally fly within their homeland with the national carrier. The first domestic line opened by Sabena was between Brussels and Liege. The service was a regional one and operated by two Britten Norman Islander aircraft, which could carry eight passengers, with a speed of 137 mph (220 kph). Sabena figured that the domestic network would be profitable with an average occupancy rate of 60%. The main aim of the new network was to pick up passengers who were traveling abroad from their provincial towns and fly them to Brussels International Airport for their international connections.

In March 1971, Sabena ordered two McDonnell Douglas DC-10-30CF long-range convertible freighters for delivery in late 1973. The Belgian airline was the first scheduled operator to purchase the mixed cargo version of the DC-10. Sabena's version could accommodate 255 passengers, 24 in 1st class and 231 in economy class. The cabin would accommodate 155 passengers in the mixed passenger-cargo configuration and hold eight pallets. At the same time, in the cargo compartment, it would carry 14 containers and four cargo pallets below the cabin. This configuration was suitable for Sabena's long-haul flights, where traffic density was lower than the Brussels-New York route. The third configuration was the all-cargo version which would carry 22 pallets in the cabin section. Sabena had ordered 2 DC-10-30 passenger versions, which were delivered to the carrier at the end of 1973. The DC-10-30s would operate primarily between Europe, Japan, Canada, and other routes with low traffic density. They would also fly the North Atlantic route alongside the 747s.

In 1970, air traffic on Sabena's route network increased by 10% compared to 1969. As a result, the ton-km transported in 1970 amounted to 447.3 million against 406.6 million in 1969. Meanwhile, production increased from 682.1 million ton-km in 1969 to 742.3 million in 1970, an 8.8% rise. Consequently, the load factor rose to 60.3% in 1970 against 59.6% for the previous year. In 1970, Sabena's airliners transported 1,470,635 passengers, or 7% more than in 1969. Moreover, operating results showed a 13.6% increase in ton-km transported, from 222.1 million in 1969 to 252.2 million in 1970.

Concerning cargo, Sabena carried 186.95 million ton-km in 1970 against 176.8 million in 1969, representing a growth rate of 5.8%. As for mail, letters, and parcels represented 8.13 million ton-km in 1970, against 7.8 million in 1969, an increase of 4.3%. Also, in 1970, Sabena's aircraft covered 32.4 million miles (52.1 million km) and completed 86,462 flight hours.

The 1971 summer timetable introduced three new destinations in Dakar (Senegal), Buenos Aires, and Santiago de Chile. While in Europe, Sabena re-established the nonstop Brussels-Stockholm route. Previously, the Belgian carrier flew from Brussels to Copenhagen, where passengers had to transfer to an SAS connection for the Stockholm leg. In Africa, on the Brussels-Kinshasa route, the oldest of the Belgian carrier's long-haul routes, Sabena put the Boeing 747 into service in the summer of 1971, with three weekly flights. The flight time was reduced by half an hour on that route, as it was on the New York service, thanks to the 747's speed. To take full advantage of the summer tourist season, from June 15 to September 15, Sabena configured the Jumbo with accommodation for 361 passengers, with no cargo space in the cabin. From September 16, however, the interior was arranged in the Combi layout. Sabena fitted out the cabin with 247 seats, with a section at the back of the airliner for nine containers.

It was that time again; the Annual Meeting of shareholders for Sabena was held on Tuesday, July 20, in Brussels, led by the Chairman of the Board of Directors, Mr. van Houtte. In 1970, the turnover of Sabena reached BFR 9 billion, and the result was an improvement of BFR 117 million compared to 1969. In the air traffic results of 1970, Sabena increased its production by 8.8% and its traffic by 10%. As a result, the carrier's load factor increased from 59.6% in 1969 to 60.3% in 1970. Additionally, the Belgian airline transported 1,470,635 passengers, a handy rise of 13.6%. Cargo and mail traffic increased 5.8% and 4.3%, respectively.

In a critical time for the Belgian carrier in 1971, Sabena decided to restructure the company after being in financial malaise over recent years. Its goal was to improve the company's profitability and finally end the era of deficits. Many thought that the airline's restructuring was farcical, as the carrier had opened more routes to overseas destinations in recent years. The Belgian airline did open a third flight to East Asia and routes in West Africa, and two new lines to Buenos Aires and Santiago de Chile in 1971; however, Sabena claimed that they were necessary for its route network. The criticism of the carrier was not warranted because Sabena did start culling some routes on several European destinations, which were unprofitable. In addition, Sabena performed a similar rationalization process on the company's representative, sales offices abroad, and certain other activities. The Belgian company wanted desperately to increase its share capital, as in 1971, it was only receiving BFR 750 million from the State, which was an insufficient amount in the 1970s. Consequently, Sabena's Board of Directors submitted proposals to the Belgian Minister of Communications for a fundamental overhaul of the Company's financial structure. A governmental working group was examining those proposals.

Away from the problems of the air traffic figures, the restructuring, and the demanding financial situation, Sabena announced some exciting news for a change. Since July 1946, when Sabena operated its first non-scheduled flights between Europe and America, the planes of the Belgian company had completed 30,000 crossings of the Atlantic. Then in June 1947, Sabena officially inaugurated its Brussels-New York service, using a Douglas DC-4, which made the transatlantic journey in 21 hours. After that, the Big Apple became connected to Brussels with Sabena's 707s and the 747 Jumbos, which shortened the time of crossing the Atlantic to only 8 hours. After New York, the carrier added more destinations in the North and South American sectors. First, there was Montreal, the French paradise in Canada, which Sabena added in 1957. Then came the Latin American routes to Mexico City, the city of the bullfights in 1960, and Guatemala, with its rich

Mayan culture, in 1968. Finally, Buenos Aires 'The Paris of the south', and Santiago de Chile, with its skiing high up in the majestic Andes mountains, were added in 1971. Sabena knew how to choose its destinations.

The 1971 winter timetable clearly showed that Sabena would have a busy winter that year. The airline claimed that its production target over the 1971/72 winter period would total around 348.35 million ton-km or 13% more than the previous year. Sabena was confident that in just five months, its production would be as much as its production for the whole of 1962. The new routes to South America, Buenos Aires, and Santiago de Chile, and its newest destination in Africa, Dakar, would certainly help in increasing the ton-km. Sabena projected a 28% increase in air traffic on all its African routes for 1971. Moreover, that year, Sabena was operating much larger and faster aircraft. Their use over the winter period would almost certainly ensure the goal that the Belgian company had set would be achieved.

Sabena would serve 38 locations in 19 countries across Europe over the winter of 1971. Many of those flights would be carried out in collaboration with other airlines. Regarding the number of flights available from Belgium, London remained in the top place. Fifty-one of the 79 scheduled weekly services would depart from Brussels, with the remaining 29 departing from Antwerp. There would be 20 weekly connections between Antwerp and Brussels. In the case of Liege, 17 weekly flights would be available. Flights from Liege, Antwerp, and Brussels were scheduled so that the best connections to Brussels-National airport were available. In the rest of Europe, Paris was provided with 34 weekly flights, while Milan was served with 26. The following cities were supplied with weekly flights: Frankfurt with 39, Copenhagen, Rome, Zurich, and Geneva received 21. Eindhoven 20, Amsterdam 36, Stockholm, Madrid, Luxembourg 14, Athens 9, and Barcelona, Manchester, Munich, and Stuttgart received 7. Sabena also flew to the following cities in Europe: Cologne, Lisbon, Hamburg, Alicante, Las Palmas, Palma de Mallorca, and Malaga. Sabena was in for a hectic winter.

In the winter, Sabena would serve 18 cities in 16 countries in the African sector. The important city for Sabena, Kinshasa, the capital of the Congo, was the recipient of 5 weekly Boeing 707 flights. The capital of Liberia, Monrovia, had two weekly connections from Belgium, one via Abidjan (Ivory Coast) and the other via Conakry (Guinea). Johannesburg was provided with four weekly flights instead of 3 during winter. In addition, regular services continued to be made available to Nairobi (Nigeria), Kigali (Rwanda), Bujumbura (Burundi), Dar-es-Salaam (Tanzania), and Entebbe (Uganda) in East Africa. Sabena also operated services to Abidjan, Dakar, Kano, Lagos (Nigeria), Niamey (Niger), Conakry, and Monrovia in West Africa. In North Africa, Sabena offered flights to Casablanca, Tunis, and Cairo over the winter season.

Sabena provided 12 weekly flights between Brussels and New York on the North American route network, of which seven were daily with the Boeing 747 and five weekly flights on the Boeing 707s. Montreal was served five times a week from Brussels, while Mexico City and Guatemala were offered two weekly flights. The new route to South America, which opened in 1971, was operated once a week. The route was as follows: Brussels-Dakar-Buenos Aires-Santiago de Chile.

In the six months since Sabena's 747 operations on January 8, 1971, the Jumbos transported 40,484 passengers and 2,939 tons of freight on the New York route alone. During that period, the 747s made 351 flights. In contrast, the Boeing 707s, which also provided daily flights between Belgium and the United States, transported 21,496 passengers and carried 910 tons of cargo. The two Boeing jetliners, working in tandem, flew 61,980 passengers across the Atlantic, or 52% more than during the corresponding period of 1970. The load factor for the Boeing 747s was 35.1% during the first quarter of the year. In the second quarter, it jumped to 54%. All in all, it was not a bad achievement by the Boeing siblings.

Photo 64. A classic Sabena 707-329C 'OO-SJM' at Munich International Airport in the middle of winter on January 21, 1973. Delivered on June 16, 1969. It was briefly leased to Transavia Holland in February 1976 and eventually sold to Sobelair on November 26, 1979. (Photo by Norbert Kröpfl)

Photo 65. Another classic image of a Sabena 747-129 Combi 'OO-SGB' at Palma de Mallorca Airport in June 1974. (Photo by Wilhelm Hell)

1972

The new year in 1972 did not bring much joy to Sabena's management. First, the good news, Sabena achieved a milestone in 1971 by introducing its two Boeing 747s on the Brussels-New York route. In addition, the opening of two routes to South America was another important event for the Belgian carrier. Sabena provided a weekly Brussels-Dakar-Buenos Aires-Santiago de Chile service using its Boeing 707-329Cs from April 1971. As a result, passenger traffic increased by 3% to 1.5 million passengers in 1971 compared to 1970. In addition, Sabena's cargo services in ton-km reached 210 million, or 12% more than the previous year. In total, the growth rate in 1971 was 15%. The most significant increase in air traffic was on the Brussels-New York route, which increased by an astounding 32%. This accomplishment was made possible by the additional capacity made available by the introduction of the Boeing 747 on the route. In addition, the Belgian company transported a total of 512,978,281 ton-km against 447,280,581 in 1970. The increases in 1971 were admirable, and Sabena's air traffic numbers improved almost every year. However, the increase in traffic did not translate into improved financial results for the Belgian carrier.

Sabena was constantly facing persistent difficulties in its financial burden and capital structure. As a result, the Belgian airline suffered a loss in 1971, even more, significant than the heavy loss in 1970, of BFR 565 million. Sabena's turnover in 1971 was BFR 1.197 billion, a satisfactory result; however, the significant increase in operating expenses, which rose by 14.8% to BFR 1.226 billion in 1971, ravaged Sabena. As a result, the company's gross operating profit fell by BFR 29 million, or 4%. Thus, the significant increase in operating expenses in 1971 was BFR 845 million higher than in 1970. The Belgian carrier's financial pressures, such as constant declines in unit revenue, continuous increases in production costs, and the severe effects of the early 1970s monetary crisis, significantly impacted Sabena's bottom line. Undoubtedly, most of IATA's scheduled carriers faced significant financial and economic difficulties in the early 1970s, which would take time to resolve. Still, Sabena's problems were far more severe than those of the other carriers.

In the previous 12 years, while many of those IATA airlines were posting profits, Sabena remained heavily in the red. From 1960 up to 1972, Sabena posted continuous losses that amounted to an approximately BFR 5.23 billion. Although the Belgian State took up those deficits, the situation could not go on like that forever. One of the reasons for the enormous losses over the years was the number of staff at Sabena. There were just far too many employees working for the airline. Another reason was the loss of the Belgian Congo, which became independent in 1960. Until that year, Sabena had a monopoly on the Belgian-Congo route, and the carrier also had an extensive domestic network in Congo. The ending of those services after Congo's independence dealt a substantial financial blow to Sabena. The third reason was the maintenance of Sabena's helicopter operations, which started in 1953. Sabena's helicopter services were far too expensive to maintain and should have been stopped very early in their operations instead of 1966, which was far too late. The final reason was the stiff competition Sabena faced from the charter airlines, causing the scheduled carrier headaches. As a result, its management announced that it would start a hard fight against all charter companies. These factors played a part in ensuring that Sabena would be a loss-making airline for many years to come.

In response to Sabena's dire economic situation in 1972, its management implemented a series of rationalization measures. These measures included reducing the workforce, discontinuing unprofitable routes, and selling some of the fleet's aircraft. Sabena intended to lay off 350 of its 10,000 employees. Early retirement and natural attrition were to be used to achieve the staff reductions. In addition, Sabena announced that it would sell four Caravelle jetliners, two to its charter subsidiary Sobelair and the other two on the open market. Furthermore, the company announced the abandonment of plans to purchase more 747 jetliners. As for the Concorde, it was the end of the road for the supersonic jetliner. Sabena's

Managing Director, Gaston Dieu, stated that Sabena "cannot afford a Concorde" given the airline's precarious financial position. Mr. Dieu also said that Sabena's 707s would be used on shorter, busier European routes like London, Frankfurt, and Nice. He added that Sabena would also shut down its services to Alicante, Budapest, Hamburg, Manchester, Palma, Jeddah, and Istanbul. The company's management hoped to halve its losses with the reorganization plan the following year.

Sabena's winter timetable for 1972 introduced the inclusion of a new stopover on the Kinshasa route, Douala, Cameroon's capital, inaugurated on December 1, 1972. In the winter of 1972/73, Sabena's aircraft were scheduled to fly 14,280 flights, totaling 364.3 million ton-kilometers. In Europe, for the first time, Boeing 707s would fly to Malaga and Boeing 727s to Munich. In addition, Sabena provided many other flights in the winter of 1972. For example, London had an astounding 75 weekly flights to Belgium. Fifty-one of those flights were between London and Brussels, while 24 were between London and Antwerp. Paris and Frankfurt were served by 34 weekly flights each, Milan and Amsterdam by 26, and Copenhagen, Rome, Zurich, and Geneva by 21.

Sabena's number of countries serviced dropped from 50 to 48 in 1972. The decrease was caused by the removal of several cities from the carrier's European network as part of the reorganization plan. Sabena offered flights to 19 cities and 18 countries on the African continent. Kinshasa was served with five weekly flights on Boeing 747s and 707s, three of which were nonstop. Further south, Johannesburg was served with four weekly flights. Other regular services to Africa were as follows: Nairobi, Kigali, Bujumbura, Dar-es-Salaam, Entebbe, Abidjan, Dakar, Kano, Lagos, Niamey, Conakry and Monrovia. Meanwhile, regular flights connected the North African cities of Casablanca, Tunis, and Cairo to Brussels.

In December, the Belgian airline revealed that the company would make some changes for its 50th anniversary in 1973. 1) Beginning April 1, 1973, Sabena's air hostesses would receive a new uniform. 2) The curvy 'S' on the planes' tails would be transformed into a stylized 'S' encircled by a white circle. The new emblem started to appear on Sabena's planes in 1973. 3) An issue of Sabena's magazine, Revue, would go through the various stages in the development of Sabena over the years. The issue would also be published in the form of a commemorative brochure. 4) An essay competition would be organized in 1973 in both the French and Dutch speaking final year high school classes, with the topic: ' Sabena, a large company in a small country.' The essay competition winners would receive free air tickets to travel to Mexico and stay there for eight days. 5) On the day of the official 50th Anniversary, May 23, 1973, Sabena passengers would be presented with roses at Brussels-National airport. The roses would bear the name 'Miss Sabena.' 6) The new Sabena electronic reservation center would be inaugurated in January. The center would make it possible to reserve seats onboard the airline's planes in a fraction of a second on departure from all significant stopovers on the network. Sabena named the service 'SAPHIR,' which stood for 'SABENA's Automated Passengers and Hotel Integrated Reservations.' 7) Sabena would release a calendar recalling the milestones in the history of the Belgian company. 8) The Belgian painter Rondas would paint characteristic images of Belgian cities, which would be reproduced in 1973 in posters that could be exhibited together or separately. 9) The Belgian Postal Commission would issue a stamp commemorating the first 50 years of Sabena. Amid its financial woes and misfortunes, Sabena's management and staff would have some reasons to smile about in 1973.

Regarding the 747s in 1972, Sabena was the first airline worldwide to modify its two Boeing 747s. Sabena asked Boeing to install a cargo door on the left side of the aircraft's fuselage, just behind the wings. The cargo door would provide access to the main deck. It would thus allow the company to substantially increase its air cargo transport capacity in the low season when fewer passengers are traveling. The maximum tonnage of freight on board would depend on the passenger/freight layout chosen according to requirements and may reach more than 90 tons. Once the cargo doors were in place, Sabena's Boeing 747s would be able to transport containers and pallets 6 meters long and 1.8 meters high.

Photo 66. What if? A model of the Concorde in Sabena colors on display at the Royal Military Museum in Brussels. Sabena's management made the right decision to cancel the orders for the 2 Concordes. Sabena just could not afford them (Photo by Clemens Vasters via Wikimedia)

The Sabena fleet in May 1972 comprised the following aircraft: 2 Boeing 747-129, 6 707-329, 6 707-329C, 2 727-29, 3 727-29QC, 7 Caravelle, 2 Fokker F.27. Light aircraft: 9 Cessna 310, 6 SIAI-Marchetti SF.260. There were also 2 DC-10-30CF on order. The option for the 2 Concordes was taken off the table.

1973

The big event of 1973 was the 50th anniversary of Sabena. However, another major event that was more important for Sabena regarding its flight operations was the introduction of 'Sabena's Automated Passengers and Hotel Integrated Reservations' or SAPHIR. The reservations system came into operation on January 15, when passengers of the Belgian company could reserve their seats in a fraction of a second. Apart from securing a seat on a flight, the reservation of hotel rooms and rental cars were among the other possibilities the new technological marvel offered. Any Sabena passenger could make reservations right up to the last minute. SAPHIR would improve the airline's load factor by around 1%, translating into more revenue. In addition, Sabena expected to achieve an annual supplement of around BFR 100 million from SAPHIR. Avoiding the considerable investment involved in the acquisition of an individual installation, Sabena decided to enter into an agreement with Air France, which enabled the Belgian carrier to secure the use of the Univac (Universal Automatic Computer) with the ALPHA 3 system, which the French national airline had been using for several years. The agreement concluded between the two companies allowed Sabena to manage the system itself, ensuring its management and control. The first foreign city to be connected to the system was London on February 12, while the two Belgian cities, Liege and Hasselt, were included on February 15 and 19, 1973. Throughout 1973, many more connections were made on the following stations: Brussels National Airport, New York, Montreal, Frankfurt, Manchester, Geneva, Paris,

Milan, Zürich, Amsterdam, Nice, Lille, Rome, Munich, Dusseldorf, Hamburg, Madrid, Lisbon, and Barcelona. The deal with Air France was a good one for Sabena.

In the air traffic results of 1972, Sabena carried 1,484,808 passengers made 41,525 flights, completed 82,439 flight hours and covered 30,750,979 miles (49,488,904 km). The company's production reached 907,651,801 ton-km, with 540,093,287 ton-km used. Production remained virtually unchanged, up by just 1% compared to the previous year, but the figure for ton-km used rose by 5.3%. In addition, the airline's load factor reached 59.5% compared to 57.1% in 1971. Sabena suffered a fairly significant reduction in unit revenue in 1972 due to meager promotional rates, primarily on the North Atlantic. Furthermore, competition on the Far East and East Africa routes became more intense. As a result, Sabena had to deal with a steady increase in production costs, particularly personnel costs. Aside from those setbacks, a few others included the increasing difficulty in obtaining new traffic rights (particularly to US destinations), Belgium's low attractiveness as a tourist destination, and the high financial charges. Despite the challenges, Sabena expanded its long-haul network in 1972 by opening new lines to Singapore and Douala (Cameroon). Concurrently, the company focused on the most profitable routes in its European route network. In 1972 it also started using its 747s on its main African route, Brussels-Kinshasa. Notwithstanding the improvement in its load factor, Sabena's financial results for 1972 showed that the airline recorded another loss of BFR 1.3 billion, compared to a revenue figure of just over BFR 9 billion.

In fleet-related matters, in early May, the Belgian Government was asked by Sabena to provide a loan of up to BFR 2.5 million to enable the airline to purchase 10 Boeing 737-200s, nicknamed the 'Tin Mouse' by its pilots. In July, the government decided, and it was a Yes. The Government approved the purchase so Sabena could replace its outdated airliners, 7 Caravelles and 5 727-100s, with the Boeing 737-200 type. During the search for a new airliner by Sabena, the only rival of the Boeing 737 was the French Mercure. The Belgian company decided on the Boeing 737-200 instead of the French twin jet because the 'Baby Boeing' airliner was considerably cheaper, costing BFR 212.14 million (USD 6 million). The first six were due to be delivered in 1974.

Sabena became the first airline globally to use the main cabin of its 747 airliners for handling and transporting cargo in the form of bulky loads. It contracted Boeing with the task of equipping two of its 747s in its fleet with a side-mounted cargo door that would be fitted on the left side of the fuselage. The modifications of its two 747s would be completed by March 1974. Later, the first of Sabena's 3 DC-10-30CFs, 'OO-SLA,' arrived at Brussels National airport on September 22. Sabena assigned it on the Tokyo line via the 'Polar Route' on April 5, 1974. Sabena's fleet by the end of 1973 was as follows: 2 Boeing 747, 6 707-329, 1 DC-10-30CF, 6 707-329C, 2 727-29, 3 727-29QC, 7 Caravelle, 2 Fokker F.27, Light planes: 7 Cessna 310, 9 SIAI SF-260. There was also one more DC-10-30CF on order.

1974

For the 1974 summer schedule, Sabena introduced new flights to Montevideo (Uruguay), Nassau (Bahamas), Kuala Lumpur (Malaysia), and Jakarta (Indonesia). It also reopened previous routes to Moscow and Istanbul. The Belgian airline also introduced its new DC-10 30CFs and placed them on the East Asian and African networks. As for its other recent additions, the Boeing 737-200 started operations at the beginning of June on the routes linking Brussels to Paris, Nice, and Lourdes. Sabena served 19 cities during the summer of 1974 with its 737-200s. As for the Boeing 747s, they were equipped with the new cargo door and put into operation on the Brussels-New York line. In Europe, 35 flights were provided on the Brussels-Paris route.

Sabena's air traffic performance in 1973 improved. The Belgian airline carried 1,649,916 passengers, 11.1% more than the previous year. While its production amounted to 946.6 million ton-km, an increase of 4.3%

Photo 67. 747-129 'OO-SGB' taxiing on a nice day at Brussels Airport. (Photo from 'www.douglasdc8.com')

Photo 68. Sabena's first DC-10-30CF, OO-SLA'. Delivered on September 22, 1973, it stayed with Sabena until December 8, 1994, when it was sold to World Airways Cargo. (Photo by Udo K. Haafke)

in 1973 compared to 1972. The ton-km used reached 635.5 million ton-km or 17.7% more than in 1972, resulting in a load factor of 67.1%, a handy 7.6% rise over the previous year. Concerning its route networks, the increase in traffic was 1.9% in Europe and the Near East, an outstanding 51.2% in East Asia, 11.9% on the North American routes, and 6.6% in Africa. As a matter of interest, the growing share of chartered flights in the air traffic results of Sabena in 1973 was 7.3% compared to 3.9% in 1972. That was due to the greater share taken by the Belgian national airline in long-haul charter activities. However, those results were unfavorably influenced by the monetary crisis, which harmed the airline's revenues, and the worldwide energy crisis, which negatively impacted many industries in the early 1970s, including the airline industry.

Later, Sabena's financial report was released, and the news was not good again. The Belgian carrier closed 1973 with a loss of BFR 759.8 million, compared to the BFR 1.3 billion in 1972. Nevertheless, Sabena's management stated that the loss in 1973 was not as bad as in 1972, and it constituted an improvement. It was like saying, I have a large hole on the left side of my pants, but I have a smaller hole on the right side. The Belgian public, whose taxes were used in supporting Sabena, must have been bored of hearing that their national carrier was constantly in the red. As for the numbers, in 1973 expenditure increased by 5%. Staff costs were extremely high, as they exceeded BFR 5 billion, an increase of 8% compared to 1972 despite a 3% reduction in the workforce. Regarding operating revenue, it increased by 10.9% compared to 1972. Sabena's reasons for the loss in 1973 were the adverse economic situation characterized by monetary instability aggravated by persistent inflation and the acute fuel crisis. The company also highlighted that its results would have been better if the carrier had not suffered the unfavorable impact of certain uncontrollable events, such as the strikes by the Belgian air traffic controllers.

In response to Sabena's financial perils, its Managing Director, Gaston Dieu, stated in late 1974 that Sabena's profit-earning capacity had deteriorated significantly over the past few months. That was due to the increase in fuel prices, which tripled in a matter of months, and the acceleration of inflation caused by the energy crisis in the early to mid-1970s. Mr. Dieu stated that all European countries were experiencing the same fundamental issues, with airlines being particularly hard hit. He went on to say that fuel costs accounted for one-fifth of the operating costs of international air travel. Furthermore, aviation was highly vulnerable due to its labor-intensive nature. Then, he admitted that Sabena's ills were not all due to external events. Even after the arrival of new airliners helped increase the airline's load factors, Sabena's costs remained deplorably high compared to those of its competitors. That was mainly true in the areas of salaries and financial overheads.

Another factor that dented Sabena's bottom line was the Belgian charter companies undercutting the national carrier. In 1972, Sabena's management stated that it would start a hard fight against all charter companies. In late 1974, the Belgian national airline took the decisive step to shut down all other Belgian charter airlines competing with its own charter company Sobelair within one year. Sabena and its subsidiary Sobelair signed a 50-year contract with the three major Belgian tour operators, Airtour, SunComfort, and Cextrair, to control the charter market and push its competitors back. The agreement stipulated that Sobelair would fly at all times and under all circumstances for the tour operators with fares that would be 7.5% lower than the lowest offer of Sobelair's competition. In addition, the tour operators would also bear half the printing costs of the luxury holiday brochures. The contract also included a very advantageous arrangement for the tour operators; Sabena and Sobelair would always utilize the aircraft type that the tour operators preferred. Thus the tour operators would not face the problem of having more passengers for a flight than there were seats on the aircraft. The tour companies were also freed from paying extra costs for a chartered airliner that was too large but was less than half full. The contract between Sabena and Sobelair with the Belgian tour groups ensured that funds would flow back to Sabena through subsidies from the state treasury. The agreement nullified any chance of fair competition for the Belgian charter airlines such as

Delta Air Transport (DAT) and Trans European Air (TEA), which would be priced out of the market. Sabena wielded the sword and cut through the Belgian charter airlines' lifelines.

The winter timetable in 1974 had a sad note to it. Sabena was withdrawing the Caravelles, and the new Boeing 737-200s would start to take over the flights that were previously the domain of the Caravelles. The 737-200s would fly to 19 cities in Europe in the winter of 1974/75. Sabena also resumed its second flight to South America, following a new route going via Casablanca, Dakar, and Montevideo and reaching its final destination of Santiago de Chile. Sabena's aircraft would operate 12,442 flights during the winter, a 1% increase over the 1973/1974 winter season. They would complete 32,011 flight hours and cover 12,199,380 miles (19,633,000 km), a 2% rise from the previous year. The number of flights and flight hours were practically unchanged compared to the previous winter. Thus, the increases of 1% and 2% were due to the extra capacity of Sabena's aircraft on the carrier's long-haul routes. In October, Sabena ordered a third DC-10-30CF. The aircraft was due to be delivered in October 1975. Sabena's fleet in 1974 consisted of the following aircraft: 2 Boeing 747-100C, 2 DC- 10-30CF, 6 707-320, 6 707-320C, 2 727-100, 3 727-100QC, 5 Caravelle, 1 Fokker F-27. Light aircraft: 7 Cessna 310, 9 SIAI-Marchetti SF260. Apart from the third DC-10CF, Sabena had 10 Boeing 737-200s on order.

In December 1974, Sabena received some excellent news. According to IATA (International Air Transport Association) data, the Belgian airline ranked 15th out of 113 member airlines in passenger-kilometers flown internationally. Regarding cargo transport, the Belgian airline was ranked 12th. Also, Sabena was ranked 16th for the number of kilometers traveled in scheduled international services and 13th for the number of international flight departures performed by its airliners. Furthermore, Sabena was ranked 26th in personnel numbers, with 9,845 persons on staff. All IATA carriers had an average of 77.5 passengers per plane, whereas Sabena had an average of 86. Meanwhile, the average passenger distance traveled by all IATA members was 914 miles (1,471 km), whereas Sabena's was 1,586 miles (2,553 km). IATA member airlines transported 229 million passengers and carried almost 3 million tons of cargo in 1973.

1975

At the beginning of 1975, Sabena became involved in a dispute with the US. The issue was an American request for Sabena to limit its flights across the Atlantic. Sabena was unhappy with the request and stated it was reluctant to drastically cut its flights on the North Atlantic route. The Belgian Ministry of Transport declared that the American proposal violated the 30-year-old aviation agreement between Belgium and the United States. Consultations with the Americans had started in Brussels in mid-January. A spokesman for the Belgian Ministry of Transport said that the US request to reduce Sabena's transatlantic air capacity was to help its own US companies. Pan American World Airways and Trans World Airlines (TWA) were all suffering significant losses on the route. The American side expected a 13% drop in transatlantic flights in 1975, which would affect all carriers flying the route. Apart from the request for Sabena to curtail its flights, the Americans believed that Sabena should also reduce the number of its passengers that were not Belgian citizens down to 4% on all its North Atlantic flights. For Sabena, that was a ridiculous demand. It meant Sabena would cut in half all its transatlantic flights if it were to do that. The Belgian Ministry of Transport said that Sabena had already reduced its seating capacity on the New York route by 27% in 1974.

Photo 69. Sabena's first 737-229, 'OO-SDA,' at Salzburg Airport on a wintry day on December 20, 1975. Delivered on April 26, 1974, and stayed with Sabena until March 8, 1999. (Photo by Michael Bernhard)

Photo 70. 737-229 'OO-SDA' landing at Brussels Airport. (Photo from 'www.douglasdc8.com')

Photo 71. 737-229, 'OO-SDE,' about to land. It was delivered to Sabena on June 20, 1974, and was utilized until April 6, 1999, when it was withdrawn from use. (Photo by Eduard Marmet).

Photo 72. 'OO-SDA' on final at Athens in July 1983. (Photo by Richard Vandervord)

Photo 73. Sabena 737-229, 'OO-SDF,' at Munich Riem in January 1976. Delivered to Sabena on July 21, 1974, and withdrawn from use on Oct 21, 1999. (Photo by Norbert Kröpfl)

Photo 74. 'OO-SDJ' on final approach to Munich, on May 24, 1979. Delivered to Sabena on March 10, 1975, and sold to European Air Charter on July 26, 2000. (Photo by Stefan Röhrich).

Photo 75. 737-229, 'OO-SDC,' coming in for a landing. Delivered on May 13, 1974. It was in operation with Sabena until April 4, 1988. Afterward, it was leased by several carriers in Europe, Africa, and South America. It was eventually broken up on January 9, 2007. (Photo by Udo K. Haafke)

The discussions dragged on over the ensuing months, with both sides disagreeing. The Belgian minister for telecommunications stated that Belgium was not in favor of a proposal by Pan American Airways for Sabena to reduce the number of weekly transatlantic flights from 14 to 7. The minister said that Sabena could not be blamed for having a more significant share in traffic on the North Atlantic route, as Belgium only had landing rights in one city in the US; New York. Previously, there were negotiations between Belgium and the US on landing rights to Chicago for Sabena. However, the US side rejected that request faster than a bolt of lightning. Eventually, after many discussions, an Aviation Air Transport Services Agreement was finally signed in Brussels on October 24, 1977, by the USA and Belgium. As a result, Sabena was finally authorized to operate flights to New York and Atlanta, and three other cities.

In February, Sabena was technically bankrupt after an unbearable loss of BFR 2.48 billion in 1974. The share capital was only BFR 750 million. The company attributed the severe loss to the usual suspects, the increasing oil prices, and higher personnel costs. According to Belgian law, companies that showed more than three-quarters of their share capital as a loss were obligated to convene their board of directors, who had the task of proposing the company's liquidation to the shareholders. However, seeing as Sabena was 90% government-owned, the state would take over the deficit and increase the airline's share capital. The poor Belgian taxpayers had to foot the bill once again. The news was not as bad as the carrier's financials in the air traffic results. In 1974, Sabena posted a 12% increase over the 1973 air traffic total of 711 million ton-km. Also, with the introduction of 2 DC-10-30s and 6 737-229s in the fleet in1974, Sabena increased its capacity by 20%.

Following Sabena's disastrous loss in 1974, the Belgian government moved on Sabena's board members. Six of the airline's fifteen government-appointed board members were replaced at the airline's annual general meeting in 1975. Furthermore, the airline was now led by a new chairman, Gaston Coppee of the Belgian Government's finance department. Sabena's management would now be subject to a Government commissioner, who would have a veto over all managerial decisions. In March, exploratory talks on a merger were held between Sabena, the Dutch national carrier, KLM, and Luxair, the Luxembourgian national airline. The idea for a union originated with the European Economic Community (EEC) transport ministers, who proposed a joint Benelux airline at a meeting in Paris in 1974. The merger idea arose after Sabena and KLM encountered difficulties with American proposals to limit their landing rights in the United States. The merger was proposed in good faith, but the road ahead was riddled with impassable obstacles.

To begin with, only Luxair, the smallest of the three Benelux airlines, was profitable. In 1975, KLM and Sabena were two of Europe's least thriving airlines, as they were under severe financial pressure. Second, in the case of Sabena, it had been losing money for 13 years in a row. However, because Sabena was a majority-owned state corporation, the Belgian government always made up for Sabena's losses. On the other hand, KLM was a private company, even though the Dutch government owned a majority stake in the airline. Third, the fleets of KLM and Sabena were as different as chalk and cheese. Sabena's fleet mainly consisted of Boeing airliners, while KLM's fleet comprised McDonnell Douglas aircraft. Finally, Sabena was part of the ATLAS group, which included Air France, Lufthansa, and Alitalia. Conversely, KLM was a member of the KSSU group, which included Swissair, UTA, and SAS. The odds for a successful union between the three Benelux countries would not be as easy as many thought.

Figures released in May 1975 showed that Sabena had a higher load factor than some other large carriers, such as TWA, Pan Am, Alitalia, and British Airways. It was an excellent result for the Belgian national airline. The list of the load factors was as follows: 1 Air Canada 64.9%, 2 Air France 63.4%, 3 Lufthansa, 61.7%, 4 SAS 60%, 5 CP Air, 62.5%, 6 Sabena 62%, 7 British Airways 58.1%, 8 Swissair 57.6%, 9 Iberia, 54.5%, 10 Pan Am. 54%, 11 KLM 52.9%, 12 TWA 50.3%, and 13 Alitalia at 49.4%.

Regarding routes, Sabena reopened the Brussels-Manchester line in April 1975, after it was suspended. Sabena provided five weekly flights to Manchester, from Monday to Friday. Additionally. Sabena introduced its new DC-10-30CFs on the Brussels-Moscow-Tokyo and the Brussels-Abu Dhabi routes. The last line was added to Sabena's route network in 1975/76. Also, in winter, from November 1, the Paris-Brussels line was operated jointly by Air France and Sabena with three daily flights. The service was operated by aircraft from both airlines.

In June, Sabena announced that four of the 15 Boeing 737-229s that it had ordered would be the QC model, which was designed to be transformed into a cargo carrier within 30 minutes. The first of the 737-229QCs, OO-SDJ, was delivered to the airline on March 10, 1975, and was equipped with a door 3.40 meters wide by 2.10 meters high to allow for palletized loads on the upper deck. Moreover, in June, Sabena ordered another 5 Boeing 737-229s for delivery in the second half of 1975. Seven of the original batch of 10 737-229s were delivered to Sabena in 1975. Sabena's fleet in early 1975, was as follows: 2 Boeing 747-100, 2 DC-10- 30CF, 4 Boeing 707-329, 6 707-329C, 3 727-100C, 3 Caravelle, 6 737-200. Light aircraft: 7 Cessna 310, 9 SiaiMarchetti SF.260. It had placed orders on 1 DC-10-30CF, 5 737-229, and 4 737-229QC.

1976

In 1976, a report was released on Sabena by McKinsey & Company, an American management consulting firm, one of the largest in the world. The report concluded that Sabena's only chances of financial survival lay in a merger with the other Benelux airlines, KLM and Luxair, like the one that took place with the three Scandinavian nations' airlines from which SAS emerged. After the Benelux merger, costs would be limited by implementing many staff reductions. The report estimated that consolidation would save the three airlines BFR 3.7 billion in the first year alone. However, there were risks involved with that recommendation causing intense political turmoil in Belgium, a nation of French and Dutch speakers, who at times were at odds on national issues. In Sabena's case, the French-speaking circles opposed Sabena losing its identity because it was an airline with French Belgian origins. The French speakers in Belgium feared that, in the event of a merger, Sabena would be dominated by the larger Dutch national airline, KLM, and indirectly by the Belgian Dutch speakers. In addition, the fact that many of Sabena's employees would lose their jobs in a merger was a significant issue in Belgium. The unions in Belgium were influential in the 1970s, as they were in many European countries. The Belgian Minister of Transport, Job Chabert, tried to appease the unions by stating that no Sabena staff member would be dismissed due to a collaboration or merger. The other obstacle was the financial malaise that had engulfed Sabena and KLM. Sabena and KLM's combined deficits in 1975 were between BFR 3.5 and 4 billion, with Sabena accounting for the lion's share of the loss, which was nearly BFR 2.5 billion. Sabena's losses over the previous 15 years totaled approximately BFR 10 billion. The Belgian carrier's high staff numbers and an unprofitable route network that focused too much on former Belgian colonies in Africa were two significant reasons for the losses. Many other opponents stated that even if the three Benelux airlines merged, they would still be significantly smaller in size and funds than giant carriers like Air France or Lufthansa. The opponents did not believe any merger would result in financial bliss. In the end, the union of the three Benelux airlines was never going to happen. The collapse was based on historical reasons between the two nations of Belgium and the Netherlands. The negotiations failed primarily because the French-speaking Walloons refused to participate in such plans with the Dutch. The other reason was that even though staff numbers did not differ significantly, KLM was two and a half times larger than Sabena at the time of the merger negotiations. The Belgians feared, and rightly so that KLM would be like an overlord over Sabena. Moreover, the Belgians believed major reorganizations would be required due to the merger, predominantly in Brussels. Thus, the two airlines turned around and returned to the way they were. Financial losses or not, many Belgians were glad the merger talks had failed.

The Belgian carrier expected a 14% increase in its air traffic results for 1975. However, that was not the case as its total results for 1975 were practically comparable to 1974. Sabena transported 719.16 million ton km in 1975, which was barely 1% more than in 1974 at 711.83 million ton-km. The carrier's supply went from 1.14 billion ton-km in 1974 to 1.2 billion ton-km in 1975, a rise of 5.3%. That resulted in a fall of 2.5% in the load factor, which dropped to 60.1%. The increase in supply was essentially the result of the introduction of new airliners in 1975, nine 737-229s and one DC-10-30, which offered greater capacity than the airliners they replaced in Europe and on certain long-haul routes. The number of passengers in 1975 decreased from 1,693,417 in 1974, to 1,680,648 in 1975, or -0.8%. Regarding the load factor, it went down from 60.7% in 1974 to 58.9% in 1975, which represented a decrease of 1.8%. In 1975, Sabena's aircraft, compared to 1974, flew 83,445 hours - a fall of 0.8%), traveled 53,248,442 miles (5,227,862 km) - a rise of 1.2%, and performed 37,478 flights - a drop of 2.3%. Concerning its route network, there was a clear worsening in the results on the North American network. While Sabena reduced its capacity on that route to consider the decline in traffic between America and Europe, the Belgian company still recorded a drop of 11.7 % in ton-km transported between Belgium and North America. The decrease was more marked in cargo tonnage than in the number of passengers. That meant the drop was indicative of a negative economic situation. On its East Asian routes, passenger numbers fell by 4%, while capacity increased by 11%. In Europe and the Middle East, production was up by 9.6%, compared to 1974. However, air traffic increased by only 6.4%. Over in Africa, Sabena saw an air traffic increase of more than 2.8%, with production only rising by 0.8%.

Photo 76. Sabena DC-10-30CF on the tarmac at O.R. Tambo International Airport in South Africa on May 1, 1985. Delivered on October 28, 1975, and withdrawn from use on June 13, 1992. It was later purchased by FedEx. (Photo by Michael Bernhard)

Photo 77. DC-10-30CF is seen here at London Heathrow in May 1985. Delivered on June 10, 1974. It was with Sabena until July 19, 1994, when it was withdrawn from use. It was bought by a company called 'Nippon Miniature Ballbearing ' in August 1994. (Photo by Richard Vandervord)

The company's financial results, released in July, were terrible. The 1975 deficit was BFR 2.53 billion, which was carried forward. According to Sabena, 1975 was a disaster for the entire aviation industry due to the global economic crisis, which coincided with a decline in activity and commerce. Rising inflation was much higher in Belgium than in several neighboring European countries, impacting spending, particularly on fuel. Despite the negative results for 1975, Sabena reported BFR 15.667 billion in operating income, a 17.2% increase over 1974. On the other hand, Expenditure totaled BFR 18.185 billion francs, a 19.2% increase over 1974. Among the major items of expenditure, staff costs were at the top of the list, increasing by 17.4% despite a 175-person reduction in the workforce in 1975. In addition to the cost of fuel, which continued to weigh on operating expenses, there were financial charges totaling BFR 509 million, representing a 25% increase over 1974. Regarding its route network, Sabena opened a new line to Cuba on May 4, which was a weekly service between Brussels and Havana. Consequently, Sabena was the second European airline to offer flights to Havana. The other was Iberia, the Spanish national carrier

In 1975, the company expanded its route network by adding new stopovers in Libreville (Gabon) and Abu Dhabi. It also reopened its stopovers in Tangier and Manchester and resumed service on the Antwerp-London route with Boeing 737s. Furthermore, Sabena added its 737-229C cargo jets to its cargo network and launched a cargo service to Copenhagen and Gothenburg in collaboration with SAS. The Belgian carrier also introduced new domestic services to Brussels from Kortrijk, Ghent, Charleroi, and Gosselies. In addition, Sabena took two initiatives to improve the in-flight conditions for its passengers. First, the carrier introduced 'non-smoking' areas on its planes, which pleased the non-smokers, and offered certain amenities for businessmen.

Sabena retired its last Boeing 727-129QC, 'OO-STE,' delivered to the airline on October 2, 1968, and sold it to World Airways. Thus, the last Boeing 727 in Sabena colors said its goodbyes to Brussels on April 21, 1976. The Belgian carrier also sold its last Caravelle to the French passenger and cargo charter airline, Catair. Interestingly, Sabena sold the only two medium-haul airliner types in the fleet with rear-mounted engines. Concerning new aircraft deliveries in 1976, Sabena welcomed 9 Boeing 737-200s and one DC-10-30CF to its fleet, thus completing its program to reinvigorate its fleet. Later in 1976, Sabena changed the schedule of its 747 passenger flights to incorporate the 747 Combi. Sabena significantly increased its freight capacity as a result. Before adding the 747 Combis on the New York route, Sabena's cargo services were handled by two 707 cargo jets. However, with the introduction of the 747 Combis on the vital New York route, Sabena managed to boost its weekly cargo capacity by an astounding 350%. In June 1976, Sabena introduced a new little airliner on its service to Heathrow, the Beechcraft 99. The Beech 99s operated two daily flights to Belgium, only on weekdays, with one of those flights going via Charleroi.

In early 1976, Sabena signed a code-sharing agreement with European Air Transport (EAT). EAT was established as an air taxi service in 1971 in Brussels by two pilots, Pirlot de Corbion and Dessain. It provided local and international charter flights and contract service flights. According to the terms of the deal with Sabena, EAT was to provide daily scheduled passenger services between Brussels, Eindhoven, Luxembourg, Dusseldorf, and Cologne on behalf of the Belgian national airline. EAT carried out the flights using Fairchild Swearingen Metroliner aircraft, which offered seating for 21 passengers. EAT's Metroliners had quick-conversion cargo capabilities. In addition, the airline's aircraft were painted in the colors of Sabena.

Photo 78. Sabena Beech 99, 'OO-WAY,' at Heathrow in November 1984. The Beech 99s provided two daily flights, on weekdays, to Belgium. (Photo by Richard Vandervord)

Photo 79. Beech 99, 'OO-WAZ,' on the tarmac at London Southend Airport in December 1978. (Photo by Richard Vandervord)

Photo 80. European Air transport (EAT) Swearingen SA226AT Merlin IVA, 'OO-VGD,' at Euroairport Basel-Mulhouse in August 1996. (Photo by Eduard Marmet)

In March 1976, Sabena's fleet consisted of the following aircraft: 2 Boeing 747-129, 4 707-329 6 707-329C, 3 DC-10-30CF, 2 727-129C, 11 737-229, 4 737-229QC, 2 Caravelle. Light aircraft: 7 Cessna 310, 9 SIAI Marchetti SF.260.

1977

Sabena performed somewhat acceptably in 1976. The carrier's supply in 1976 amounted to 1.28 billion ton-km, or 4% more than in 1975. The ton-km used was 784.26 million, 9.3% higher than the previous year. Consequently, the load factor improved by an acceptable 3%, rising from 58.3% in 1975 to 61.3% in 1976. Concerning passenger traffic, the number of passengers in 1976 fell by 3.4%, compared to 1975. On the other hand, the cargo figures shot up by 27.7%. Moreover, Sabena's non-scheduled cargo transportation reached the record figure of 49.9 million ton-km in 1976, an astronomical rise of 743.6% compared to 1975. Mail transport also experienced a significant increase. It rose by 14.8% in 1976. In air operation results, the number of miles traveled was 32,519,757 (52,335,477 km), almost identical to that of 1975, while the number of departures rose, by a substantial 10.3%, to 41,324.

In the financial results for 1976, Sabena did not have much luck. While most major European airlines achieved quite good results in 1976, Sabena's figures were dismal. As a result, the Belgian carrier posted a BFR 1.8 billion loss. Fortunately, the Belgian Government stated it would increase the airline's share capital from BFR 750 million to BFR 5,000 billion within the next four years.

On June 15, 1977, Sabena announced that it had signed a technical cooperation agreement with Luxair related to their respective Boeing B-737-200s. Under the agreement terms, the two carriers' agreed that their aircraft would meet the same standards. Furthermore, the two partners declared that they would also pool their operational experiences. In addition, Sabena and Luxair agreed to consult each other in the future in all areas where an extension of the cooperation would be beneficial to both parties. Luxair was inspired by Sabena's purchase of the 737-200s, and it decided to purchase its own Boeing Baby Jets after a thorough examination of that airliner.

Sabena reached a milestone in 1977. Its Brussels-New York route was 30 years old. On June 4, 1947, the route was inaugurated when one of Sabena's DC-4s made the first flight to the Big Apple. Before the inaugural flight, there were several preparatory flights conducted. The first of those flights was made on February 18, 1946, when a Belgian crew took delivery of the carrier's first DC-4 and flew it across the Atlantic to Belgium. In 1947, there was only one weekly flight to New York, but in 1977 there were nine 747 flights per week to New York, including one via Montreal. In the 30 years that Sabena had flown the Brussels-New York route, its planes transported more than 2 million passengers. Moreover, the Belgian airline also carried 190,663 tons of cargo and 6,215 tons of mail and postal parcels.

1978

After an Aviation Air Transport Services Agreement between Belgium and the USA was signed in Brussels on October 24, 1977, Sabena finally got what it wanted, a second US destination. It was the capital of the State of Georgia, Atlanta. On June 1, 1978, the new route from Brussels to Atlanta, flown by Sabena's Boeing 707-329Bs, was inaugurated. Sabena provided four weekly passenger flights and two weekly all-cargo flights. Sabena was the first European airline authorized by the US to fly to Atlanta. After three months of operations, the ton-km used reached the figure of 11,759,019, which resulted in a good load factor of 58.1%. The total ton-km offered in the first three months was 20,230,821. For the four 707 passenger flights, the load factor recorded was 58%, while for the two 707 cargo flights, the load factor reached an estimable 64.9%. Sabena was so pleased with the results of the Atlanta line that it decided to move on to the second phase of its operation. It would replace all six 707-229s on the route with the Boeing 747-229s from April 1, 1979. The 747s were the better fit for the route, as they could combine the transporting of passengers and cargo.

In April, the Belgian Ministry of Transport announced that Sabena's organizational structure would be drastically altered before the summer of 1978. The Government decided that it was time Sabena became a wholly state-owned enterprise. With that decision, the Belgian Ministry of Transport declared that Sabena would thus cease to be a public limited company from June 15, 1979. Consequently, Sabena's main shareholder, the Belgian state, requested Sabena's management to resign at the May 1978 general shareholders' meeting. That was due to the Belgian government's desire to eliminate representatives of private capital within Sabena, despite the possibility that business people would be appointed to fill some positions after the reorganization. Sabena expected to achieve positive operating results again within the next five years due to the new organizational structure. However, the Belgian airline expected to continue operating at a loss in the short term.

In October, the new Sabena chairman of the board of directors, Mr. Van Rafelghem, stated that the restructuring plan was based on four points: moderate expansion, strong sales development, spending cuts, and productivity improvements. However, he did admit that Sabena would still not be out of the red by the end of the 5-year restructuring period in 1983. In the preceding years, Sabena had suffered so many losses in billions of Belgian Francs that it would take a mammoth turnaround to bring the carrier into a favorable financial position. He added that everything was done to reorganize the company in recent months. Mr. Van Rafelghem declared that Sabena's wage costs needed to be reduced for the airline to go forward. Sabena's employees were among the highest paid in Western Europe. Despite that, the airline stated there would be no layoffs. The company planned not to recruit any new employees, as approximately 1,300 to 1,500 Sabena staff members were due to retire from the airline by 1983. Any vacancies would only be filled when strictly necessary. Regarding expanding the airline, Sabena wanted to further enhance the fleet by purchasing another 4 DC-10-30 long-haul airliners and three medium-range aircraft.

Once again, Sabena suffered a loss of BFR 2.23 billion in 1977, despite a traffic increase of 5.7%. The 1977 deficit was 49% more than in 1976, when the shortage was BFR 1.5 billion. Regrettably, Sabena (and Air France) was the only major airline in Western Europe to post a loss in 1977. The loss was reduced by BFR 1 billion with the subsidy from the Belgian Government.

In June, Sabena reached out to British Caledonian Airways (BCal) and signed an agreement to work closely with the British airline. Sabena called the cooperation agreement with BCal 'one of the most comprehensive agreements between airlines ever reached'. The cooperation mainly covered the field of marketing. The partnership pact started on July 1. The route networks of Sabena and BCal would include up to three joint flights per day between Brussels and London-Gatwick. As a result of the agreement, Sabena would be able to offer flights to South America, Africa, and Asia due to BCal flying to many exotic destinations. BCal's South American flights were to Brazil, Argentina, Ecuador, Chile, Venezuela, Colombia, and Peru. Africa: Algeria, Tunisia, Morocco, Libya, Angola, Ghana, Nigeria, Gambia, Ivory Coast, Liberia, Senegal, Sierra Leone, and Asia: Dubai, Hong Kong.

Strong links between Belgium and London characterized Sabena's 1978 winter timetable, with 96 weekly flights offered jointly by Sabena and British Airways. The British capital could be reached 16 times a day from five Belgian cities: Brussels, Antwerp, Liege, Charleroi, and Ostend. In addition to the 96 weekly flights to the UK, there were also 26 flights operated jointly by Sabena and Air France between Brussels and Paris. Moreover, Sabena collaborated with several other European airlines, which enabled it to offer 47 flights per week to Amsterdam, 28 to Frankfurt, Geneva, Milan, 27 to Copenhagen and Zurich, and 21 to Rome. Other services from Brussels were available to the following European cities: Eindhoven, Stockholm, Madrid, Athens, Barcelona, Munich, Cologne, Bonn, Stuttgart, Manchester, Lisbon, Basel, Dusseldorf, Malaga, Warsaw, Moscow, Istanbul, and Vienna.

In the Middle East, Sabena offered two weekly flights to Doha, Qatar, starting in the winter of 1978. There were ten other destinations in the Middle East for Sabena. They included the following cities: Beirut, Baghdad, Cairo, Dhahran, Doha, Abu Dhabi, Dubai, Tel Aviv, and Tehran. On the North Atlantic lines, nine Boeing 747 flights to New York, three to Montreal, and two to Mexico City were provided each week. Sabena and Air Canada jointly operated the Brussels-Montreal route. The newest city in Sabena's route network, Atlanta, was served six times a week (including two times by 707 329C cargo jets). Sabena's DC-10-30CFs were quite busy in the winter of 1978. Sabena assigned them to the following routes: Brussels to Tokyo, Bombay, Bangkok, Kuala Lumpur, Singapore, Manila, and Jakarta. In Africa, Sabena used its DC-10s on the southern routes to Kinshasa and Johannesburg and cities in western Africa such as Libreville and Lagos. Other African destinations made available by Sabena in 1978 were: Nairobi, Kigali, Bujumbura, Dar-es-Salaam, Entebbe, Abidjan, Dakar, Kano, Conakry, Douala, Monrovia, Casablanca, Tangier, Tunis, Algiers and Cairo.

In fleet matters, Sabena ordered one 737-229 for delivery in May 1979, while its subsidiary, Sobelair, ordered two of the same type. All three were due to be delivered in September 1979 and April 1980. The carrier's fleet in early 1978 was as follows: 2 747-129, 1 707-320, 5 707-320C, 3 DC-10-30CF, 10 737-229, 4 737-229C, 1 Caravelle. Light aircraft: 6 Cessna 310, 9 SF.260.

Photo 81. Sabena 737-229C 'OO-SDK' on the tarmac in the twilight hours of a late afternoon. Delivered on March 22, 1975, and sold to European Air Charter on November 30, 1999. (Photo by Eduard Marmet)

Photo 82. DC-10-30CF, 'OO-SLD' basking in the sun. Delivered to Sabena on July 15, 1980, and was eventually sold to FedEx on February 20, 1992. (Photo by Udo K. Haafke)

1979

In February, Sabena's new President and General Manager, Carlos van Rafelghem, expressed his view that the Belgian national airline would be able to reduce its deficit over the next few years significantly. He was somewhat confident that by 1980 the debt could be reduced hugely to BFR 1.3 billion. Later, in July, at the Annual General Meeting of Shareholders, Sabena's operational and financial results for 1978 were revealed. Sabena had been subjecting its shareholders and, sadly, the Belgian taxpayers to excessive losses for more than a decade. To the attendees' surprise, however, the financial results for 1978 were considered encouraging, even though 1978 was a year of transition for Sabena, with significant changes to the statutes, followed by the establishment of new structures. The gross financial results showed an improvement of some BFR 500 million since they went from BFR 2.38 billion in 1977 to BFR 1.55 billion in 1978. That meant that the statutory intervention by the Belgian State was much less than in 1977. Concerning traffic results, Sabena announced, for the first time in its history, it passed the 2 million passenger milestone at 2,027,128. The airline sold 884 million ton-km in 1978, 11.9% more than in 1977. Moreover, production increased by 7.4% to reach 1.403 billion ton-km offered. Since the increase in Sabena's output was lower than its sales, the load factor stood at 63%, an upswing of 2.6% compared to the previous year. It looked like there might be some light at the end of the tunnel for Sabena's poor shareholders.

Concerning cargo, Sabena handled more than 137,000 tons of goods at Brussels airport in 1978, which increased by 8.8% compared to 1977. The total for 1978 was Sabena's cargo and cargo processed for 33 other airlines. The Belgian carrier's air freight transport services were developing quite quickly. So quickly, it was apparent to Sabena's management that the cargo sector represented almost half of the airline's activity, expressed in ton-km. Given the expected further growth of cargo-related services, Sabena adapted its air freight terminals to handle pallet loads and used computer technology to process the data for all the cargo. The construction of a newer and bigger air cargo terminal that would handle 300,000 tons of goods per year was to be completed by 1980.

In the summer timetable of 1979, Sabena added a new destination, Dublin, Ireland. The inaugural flight to that island nation was on a Boeing 737-229 and took place on April 2. Atlanta, the newest US destination for Sabena, was being serviced with the Belgian carrier's Boeing 707s. From the 1979 summer season onward, the Georgian state capital was also provided with Boeing 747 fights. Four weekly flights were planned on the Jumbos, which would increase capacity on the route by offering double the number of seats between Belgium and the South-Eastern region of the United States. As for New York, the Big Apple was provided with nine weekly flights to Brussels, also by 747s. Montreal received five weekly flights as part of a joint operations agreement with Air Canada. Sabena would fly to the Canadian city three times a week with its 747s and two times with 707s.

Further south, the Brussels-Mexico City route was serviced with two weekly flights on Sabena's Boeing 707s. Sabena's networks in West and East Africa were restructured to improve services to several destinations. From 1979, Sabena made it possible to double the number of flights to Monrovia and Dakar and enable a direct service from Brussels to Abidjan. The services to Bujumbura and Dar-es-Salaam were also improved.

In the Asian region, the Belgian minister of transport, Mr. Chabert, and his Japanese counterpart Mr. Moriyama signed an agreement in September that allowed Sabena to provide a second polar flight to Tokyo via Anchorage. Consequently, Sabena dropped the Brussels-Tokyo flight that flew via Southern Asia.

After much speculation about the next jetliner to be purchased, Sabena announced in November that it had placed an order with Airbus Industries for three wide-body Airbus A 310s, with options on three more. Airbus managing director Roger Beteille and Sabena President Carlos Van Raflelghelm signed the contract in Brussels on Wednesday, October 31. Furthermore, the A310s were ordered with the approval of the Belgian government, which agreed to guarantee Sabena's financing. The Airbus planes would have 215 economy seats and 18 first-class seats. Sabena chose the Pratt & Whitney P&W JT9D-7R4 engines for the aircraft's power plants. The new A 310 wide-body airliners were scheduled for delivery in early 1984, and they would be assigned to operate on European and Middle Eastern routes. Sabena chose the Airbus A310 because of its low fuel consumption and fuselage compatibility with Sabena's other widebody jetliners, the DC-10s and 747s.

In the same month that the Belgian airline ordered the A310s, it placed another major order. On November 5, Sabena announced that it had decided to move forward with its plan to double its fleet of McDonnell Douglas DC-10-30s over the next few years. During a press conference, Carlos van Rafelghem announced that the fourth Sabena DC-10 would inaugurate the new Brussels-Chicago line on August 15, 1980. Moreover, Sabena stated that it would purchase a fifth DC-10 which would be put into service at the end of 1980. The carrier claimed that it had also taken an option on a sixth DC-10 after the Belgian Government approved it. The government also approved the increase of Sabena's capital from BFR 5 billion to BFR 8 billion in 3 successive year-by-year allotments starting from 1980. In early 1979, Sabena's fleet was as follows: 2 Boeing 747-100, 5 707- 320C, 3 DC-10-30CF, 11 737-200, 3 737-200C. Light aircraft: 6 Cessna 310, 9 SF.260.

Photo 83. This 737-229C, 'OO-SDK,' paints a nice picture on a brightly lit tarmac at night at Rotterdam The Hague Airport. It was delivered to Sabena on March 19, 1975, and then withdrawn from use after 24 years on November 19, 1999. (Photo from 'www.douglasdc8.com')

Trying to Achieve the Dream
1980-1989

1980

From a quiet start to the new decade, Sabena's first significant event was the ordinary general meeting of shareholders of Sabena, which was held on June 17, 1980. The results obtained by Sabena for the 1979 financial year were markedly better than those of the previous year. The company's deficit was BFR 1.8 billion, which improved by BFR 525.8 million over 1978. However, considering the statutory State intervention, which amounted to BFR 1.1 billion, the loss to be carried forward from the 1979 financial year was BFR 670.9 million, compared to the much higher amount of BFR 1.17 billion the previous year. Despite the improvement in 1979, the company was confronted with severe cash flow difficulties caused by the lack of capital.

An unfortunate event that triggered the grounding of all DC-10s worldwide was the crash of an American Airlines DC-10 after taking off from Chicago Airport on May 25, 1979. The Federal Aviation Administration began a comprehensive investigation (FAA) that affected Sabena. All airlines worldwide that had the DC-10 in their fleets saw their services disrupted for 16 days. For Sabena, that resulted in a loss of an estimated BFR 120 million in earnings. There were other effects in 1979 that caused further financial distress for Sabena. Since deregulation had taken hold in the United States, many carriers offered fares at much lower prices than what Sabena could offer, causing the Belgian airline to lose its market share. It was a dog-eat-dog world. Another issue, which was out of control, was the price of fuel, which kept going up with no sign of easing. There was also the issue of the constant depreciation of the currencies of certain countries, which caused financial instability worldwide. Moreover, some geopolitical events in 1979 also caused problems for the air carriers, such as the Iran hostage crisis. Running an airline in a world in turmoil was no easy feat for the management of Sabena.

Due to its financial difficulties, Sabena decided to implement drastic austerity measures to help it get out of its economic challenges in December. Sabena's management announced the following steps: shutting down unprofitable routes, selling fuel-inefficient airliners, and reducing staff numbers. The route reductions occurred on the flights to Southeast Asia, as Sabena's flight frequencies to that region dropped from 4 to 3 weekly services. As for fuel-hungry aircraft, all of Sabena's Boeing 707s would be removed from passenger services and be only used for cargo flights. Later, they would be partially replaced by the DC 10s and Boeing 737s. In addition, Sabena wanted to reorganize its salary scheme and reach an agreement with the unions and employee representatives on the staff reduction measure, which was the most delicate of the three cuts. Finally, Sabena's management wanted to get its staff numbers down to 9,500, from 10,100.

Sabena's route network got a boost in November 1980. A Belgian-Brazilian deal was signed in Brussels, enabling Sabena to open a regular route to Rio de Janeiro with one weekly flight. As a result, Sabena's flights from Brussels to Montevideo (Uruguay), Buenos Aires (Argentina), and Santiago (Chile) would be via Rio de Janeiro. However, in Europe, the Belgian carrier ended flights to Eindhoven in the Netherlands due to the decline in Sabena's passenger numbers due to the improved road network between Brussels and Eindhoven. Sabena was the first airline to make Eindhoven accessible by air when it started a regular helicopter service to that city in 1955.

Sabena made some changes in its 1980 winter schedule, released in November. From February 20, 1981, the airline's McDonnell Douglas DC-10-30s would fly to two west African cities, Kano (Nigeria) and Brazzaville (Republic of the Congo). Sabena assigned 2 DC-10-30s, which it had received in July and August, to those African cities. Moreover, Sabena had plans also to use the other 5 DC-10s it currently had in its fleet on other African destinations. The reason for that was Sabena wanted to replace the four remaining Boeing 707s on its African routes due to the high operating costs of those aircraft. On the North Atlantic network, Sabena stated that from February 20, 1981, it would replace its four weekly DC-10 flights from Brussels to Chicago with three weekly flights on its Boeing 747s. The 747s would also replace the DC-10s on the Brussels-Montreal-Detroit route with three weekly flights. Over in the Middle East, flight frequencies decreased on the Dhahran and Doha routes. In Europe, in addition to the Brussels-Eindhoven line, the Brussels-Antwerp line was also canceled by Sabena for the same reason as the Eindhoven one; the road network in western Europe was more modern and more convenient for the people to drive, negating the need to fly. In Ireland, the departures of the four weekly Brussels-Dublin morning flights were rescheduled for 7.00 pm departures, with the return taking place the following morning at 8.40 am.

Finally, Sabena's fleet in mid-1980 was as follows: 2 Boeing 747-129, 5 707- 329C, 3 DC-10-30CF, 11 737-229, 4 737-229C. Light aircraft: 6 Cessna 310, 9 SF.260. Also, on order: 2 DC-10-30CF, 3 Airbus A310, and 5 Embraer EMB 121 Xingu aircraft.

Photo 84. A Sabena Embraer EMB 121 Xingu (of the Belgian Aviation School), 'OO-SXE,' at Faro, Portugal in 1988.
(Photo by Pedro Aragao via Wikimedia)

1981

The new year in 1981 was no different than any other year for Sabena. The Belgian airline was deep in financial turmoil and was facing bankruptcy. The carrier more than doubled its losses in 1980 compared to 1979, with a net deficit of BFR 1.6 billion. Sabena counted on the state subsidy of about BFR 1 billion to BFR 1.5 billion in 1981. Between 1975 and 1980, the Belgian Government pumped BFR 8 billion into Sabena, and the losses kept coming. The Belgian Ministry of Transport stated there was a possibility that Sabena would be shut down for good. However, that would be unthinkable in the nation of two peoples, as 10,000 people would lose their jobs. Moreover, there would be political ramifications for the government that allowed that to happen. However, that did not stop the Belgian Finance Ministry, which informed the Belgian transport minister, Valmy Feaux, that Sabena must drastically cut its personnel costs and staff numbers to become profitable again. The Ministry suggested that the workforce be cut by 20% or 2,000 employees, and wages should be considerably cut. Sabena's wage costs were 17% higher than other European airlines such as Lufthansa, Air France, and Alitalia. Moreover, for some senior positions, Sabena was paying salaries that were 51% higher than other airlines.

A few days later, the Belgian cabinet met and decided not to give Sabena the requested BFR 1 billion, but only half BFR 500 million. The Belgian Government believed that Sabena should first ensure that its wage and production costs were under control. Furthermore, a ministerial working group told Sabena to create a remediation plan which would be examined by the cabinet for the economic, financial, political, and social consequences. Only then would Sabena receive more Government funds.

In January, meetings were held on which unprofitable routes Sabena should shut down. The decision was not made by Sabena's management but by the Belgian Government, which was the majority shareholder in Sabena. The state decided to remove the following destinations from the company's network: Mexico City, Jakarta, Tehran, Doha (Qatar), Dar es Salaam (Tanzania), and Libreville (Gabon). The Government believed that the decision to cut those destinations should improve Sabena's operating results by BFR 390 million. One other reason for removing those cities from the carrier's route network was the high cost of fuel. Sabena's oil bill rose from BFR 3.9 billion in 1979 to BFR 5.5 billion in 1980, a sharp increase of 71%.

There was no respite for Sabena. In August, the carrier took a hit from the Belgian Transport Minister, Valmy Feaux, who stated that he wanted to ease the restrictions on charter airlines significantly with the start of winter, which would ultimately limit the quasi-monopoly of the state airline Sabena in air transport services. The Belgian charter companies were to be given more leeway regarding take-off and landing rights and set their fares. Mr. Feaux expected that would result in considerable drops in fares, which were primarily dictated by Sabena. In addition to the two top Belgian charter airlines in 1981, Sabena's subsidiary Sobelair and Trans European Airlines (TEA), the smaller charter airlines such as Delta Air Transport (DAT) and European Air Transport (EAT) would all benefit from the changes.

During the financial upheaval, Sabena took out another loan in March to buy 2 Airbus A-310s for delivery in march 1984. A European banking consortium led by the French bank Credit Lyonnais granted the state-owned Belgian carrier a loan of BFR 3 billion. Sabena's fleet in mid-1981 consisted of the following aircraft: 2 747-129, 4 707-329C, 5 DC-10-30CF, 11 737-229, 4 737-229C. Light aircraft: 5 Cessna 310, 8 SF. 260, 5 EMB 121 Xingu. There were 3 Airbus A310s on order.

Photo 85. Sabena 737-229, 'OO-SDN,' at London Heathrow in March 1985. Delivered on December 12, 1975. It stayed with the airline for 25 years and was withdrawn from use on November 30, 2000. (Photo by Richard Vandervord)

Photo 86. Sabena 737-229, 'OO-SDK', on April 22, 1981. (Photo by Udo K. Haafke)

1982

In early March, the Belgian minister of transport, Herman De Croo, gave an ultimatum to Sabena's Board. He said that Sabena would have to disappear unless the airline came up with a clear savings plan within five weeks. De Croo states that he had already planned how Sabena could be closed down. The Minister believed that Sabena should save at least BFR 1 billion a year on personnel costs, meaning the carrier should reduce staff numbers and wages. De Croo was annoyed with the carrier's management, which allowed the airline to constantly make losses of around BFR 180 million, year in and year out. He believed that even though reorganization plans were put on paper five years earlier, they were not enough to stem the bleeding. He sarcastically claimed that Sabena already received BFR 500 million per year for its staff not to come to work.

Photo 86. Belgian Transport Minister Herman De Croo, castigated Sabena's management in 1982 (Photo by Michiel Hendryckx via Wikimedia)

Not long after Mr. De Croo's strong criticism of Sabena, the airline's director, Carlo van Rafelghem, announced that the carrier's board of directors had decided on an income moderation of 2% to 17% for the entire staff of Sabena. He added that the income reductions were expected to generate BFR 1 billion by 1982. Mr. van Rafelghem said there was no question of the shutting down of Sabena. The company owned many assets for that to happen. Moreover, as the airline's main shareholder, the Belgian State guaranteed to assist in 80% of Sabena's debts, which was why the carrier's capital had to be augmented. Therefore, Sabena hoped for a capital increase of BFR 18.7 billion to BFR 26.7 billion to be saved from bankruptcy. The Belgian airline's employees numbered 10,000, roughly 423 employees for each Sabena aircraft. However, director Van Rafelghem said that the staffing at Sabena should not be compared with that of other airlines because the structure was completely different, as Sabena completed work for a lot of other airline companies. With the airline's constant strikes, Sabena's management discussed pay cuts with the unions. The unions had until the end of March to respond to the agreed decisions. However, Van Rafelghem said that Sabena's staff clearly showed that they also felt strongly about keeping Sabena going and favored income moderation. Moreover, Sabena's employees supported a definitive bailout of Sabena. They wanted their airline to keep flying.

At the end of March, 68% of Sabena's staff approved introducing the austerity measures demanded by Herman De Croo in a referendum. Sabena was in a bind with the austerity measures, as it would be more difficult for the airline to obtain any additional capital increases. In voting for the austerity measures, the 9,600 staff members were committed to surrendering an average of 2% of their wages. Personnel in the higher wage brackets would lose up to 17%. The pilots were not too happy with the decision. They threatened to strike for three days from April 16 because the pilot's association did not agree with how the Belgian government and the trade unions joined together to make decisions for them.

The Belgian government was prepared to put BFR 2.7 billion into Sabena, despite Sabena's belief that the amount needed was BFR 6.5 billion. Sabena's management claimed that a third of the BFR 2.7 billion was promised in 1978 as part of a previous cash injection but did not want to force the issue because the government was unhappy with Sabena soaking up the funds. New capital accounted for the remaining BFR 1.8 billion. The Belgian carrier was also required to save BFR 450 million per year as part of the austerity measures by reducing its route network. As a result, Sabena expected to lose most of its services to Eastern European countries. Life for Sabena was never easy, but things would not get better for the airline with the austerity procedures.

Just when Sabena thought things could not get any worse, in June, the carrier released its financial results, and they were not satisfactory. Sabena reported its most significant annual loss ever in 1981. The worst thing about the result was that it was also the 25th year in succession that Sabena made a loss. That was two bad records in one year for the carrier. The troubled airline made a net loss of BFR 3.9 billion in 1981, even though its productivity increased and the average load factor increased to 65.8%. One reason for its misfortune was that the airline was under-capitalized and was forced to borrow heavily at high-interest rates. Other factors contributing to Sabena's losses were an extremely unfavorable dollar exchange rate, which adversely affected the airline's fuel costs. In addition, the Belgian airline had considerable funds tied up in Zaire, with no way of getting the money out from that country without reaching a compromise with the Zairean Government. As a result, Sabena expected its survival plan of redundancies, wage cuts, and operational cost-trimming measures to bring it into the black shortly. One can only describe those hopes as wishful thinking.

In the summer of 1982, there were only a few changes in Sabena's timetable. First, the carrier announced the reopening of its Dar-Es-Salaam line and a weekly flight to Kilimanjaro for tourists to see the majestic mountain, the highest in Africa and the highest single free-standing mountain above sea level. Those two destinations were now part of the African network of 27 cities, regularly served by Sabena's DC-10-30s. Then, in the Middle East, Sabena increased the frequency of flights between Tel Aviv and Brussels from 3 to 4 per week. The route was operated with Sabena's Boeing 737s, but one of Sabena's DC-10s operated the Sunday flight for the summer season.

There were alterations made to the timetable for the winter of 1982/83. Sabena flew to New York daily with a Boeing 747, and there was a twice-weekly service to Detroit. In addition, Sabena offered three weekly flights to Chicago, Montreal, and Atlanta. The latter's flights increased to four weekly on March 1, 1983. DC-10s operated all flights to Atlanta. Over in Africa, all of Sabena's flights to that continent continued to be performed by DC-10s for the whole winter period. Kinshasa was served with five weekly flights, while Lagos (Nigeria), Johannesburg (South Africa), Kano(Nigeria), and Abidjan (Ivory Coast) were all provided with two flights per week. Meanwhile, Douala (Ivory Coast), Libreville (Gabon), Brazzaville (Republic of the Congo), Dakar (Senegal), Monrovia (Liberia), and Conakry (Guinea) were serviced with one flight per week. In East Africa, the cities of Entebbe (Uganda), Kilimanjaro, and Dar-es-Salaam (both in Tanzania) were also provided with one weekly flight each. At the same time, Nairobi got two flights per week. In North Africa, Sabena's Boeing 737-229s flew to Algiers (Algeria) four times a week, Tunis (Tunisia), Casablanca (Morocco) three a week, and Tangier (Morocco) once a week.

On the East Asian network, Sabena's DC-10s operated all the flights. They provided two flights per week to Bombay, Bangkok, Kuala Lumpur, Singapore, and Tokyo and one flight per week to Manila. From October 12, Sabena resumed its flights to Beirut in the Near East region, with one service per week on the 737-229. The 737s also flew the Tel Aviv, Cairo, and Istanbul routes three times a week. One of the three flights departing from the Egyptian capital stopped over at Larnaca (Cyprus).

1983

There was positive news for Sabena at the beginning of 1983. Belgian commercial institutions and the Government injected BFR 4.8 billion into Sabena. By doing so, Sabena could pay off its short-term debts and some of its long-term ones. Moreover, the funds would allow Sabena to break even in 1983, the first time it would be able to make a profit since 1957. In December 1982, the Belgian Government raised Sabena's equity to BFR 9 billion, from BFR 3.4 billion. Sabena stated that it was continuously making losses in past years due to under-capitalization. Before the recent cash increase, its equity was only BFR 750 million. That forced Sabena's management to borrow money at high-interest rates from financial institutions for any significant outlays. Commercial organizations contributed about two-thirds of the latest capital infusion. Moreover, the Belgian Government's stake in Sabena fell from 98% to less than 60%. In April, there was more positive news. Sabena announced it had made a profit of BFR 23 million for the first quarter of 1983, compared with a BFR 777 million loss for the corresponding period in the previous year. It was the first time since 1957 that the Sabena had come out of the red.

Sabena's financial results released in June showed an improvement over 1981. The company's management said that the airline lowered its losses to BFR 2.15 billion in 1982, compared to BFR 3 billion in 1981. In 1983, Sabena was expected to implement stringent measures to improve its financial situation, including increasing the carrier's capital, laying off about 1,000 employees, and implementing a 17% wage cut. Following the decision to reduce Sabena's workforce by early retirement, one measure that positively influenced the airline's balance sheet was just that. The Belgian airline already spent BFR 2.5 billion funding the early retirement scheme. To save further costs, Sabena was even considering leasing new aircraft rather than purchasing them outright in the future, as was the case with Sabena's three A310s, the first of which was due to be delivered in 1984. Van Rafelghem stated that he was optimistic Sabena's financial situation would improve dramatically by the end of 1983. He was entirely correct; all was looking good by August. For the first time in two and a half decades, Sabena was on its way to posting a profit. In the first seven months of 1983, net income was in positive territory at BFR 658 million. Concerning Sabena's air traffic performance measured in sold ton-km, the carrier's passenger traffic increased by 1.6% in 1982, while the seat load factor was 64.2%.

At the end of 1983, a leasing company was established in Brussels to assist Sabena in financing its aircraft purchases. Aviafin was a Belgian company that would buy Sabena's Airbus A310s and lease them back to the carrier. Sabena had ordered three Airbus A310s, two of which were scheduled to arrive in 1984 and the third in 1986. Sabena provided 27% of the equity in Aviafin, which was founded with BFR 2.75 billion in capital. The second-largest Belgian bank, Banque Bruxelles Lambert, had the most significant financial stake in the airline. Another 11 Belgian banks, insurance companies, and credit organizations provided the remaining equity.

1984

The news was out. Sabena finally made a profit in 1983, for the first time in 25 years. The Belgian carrier posted a net profit of BFR 22 million. There were three main reasons why Sabena entered the black in 1983. There was a drastic cutback in staff wages of 17%, the airline managed to save on fuel consumption, and there were reductions in the numbers of staff, as 1,000

Photo 87. DC-10-30CF 'OO-SLC' taxiing at Brussels Airport. Delivered to Sabena on October 28, 1975. Sold to FedEx on October 27, 1992. (Photo from 'www.douglasdc8.com')

Photo 88. Sabena 737-229 at London Heathrow in May 1996. Seen here in Sabena's third and final livery. Delivered on June 18, 1974. After being leased to several airlines, it was sold to European Air Charter and re-registered 'G-CEAC' on June 11, 1999 (Photo by Richard Vandervord)

employees accepted early retirement, which resulted in Sabena having 8,762 employees at the end of 1983, compared to more than 10,000 in 1978. In addition, the airline's capital, which was increased in early 1983 by the Government from BFR 3 billion to BFR 9 billion, also helped reduce the very costly repayments of financial charges on its loans. Meanwhile, Sabena's operating results for 1983 were superb. The carrier's productivity was 1.4 billion in ton-km offered, a decrease of 3.9% compared to 1982. However, ton-km sold exceeded 984 million, or 0.8% more than in 1982. That minor increase translated to a 4.8% rise in the airline's load factor, from 66.8% to 70%. The load factor for passenger traffic was also positive at 65.3%, or 1.7% higher than the previous year. Those numbers meant that Sabena ranked amongst the top group of European airlines. Moreover, Sabena carried 1,956,886 passengers in 1983, and its planes flew a combined distance of 30,365,460 miles (48,868,472 km).

Regarding Sabena's fleet, on February 23, Sabena took delivery of its first Airbus A310-222 'OO-SCA' and received its second A310 'OO-SCB' on March 30. The two aircraft were equipped with 211 seats, 193 in economy and 18 in first-class. Initially, Sabena temporarily used the A310s on its European routes from Brussels to London, Milan, Rome, and Stockholm. Later, it permanently assigned the two Airbuses on its routes to Africa and the Middle East. Interestingly, Sabena's first A310 was the first to be delivered with a maximum take-off weight (MTOW) of 142 tons, which was 10 tons heavier than the previous A310s delivered to other airlines with an MTOW of 132 tons. The weight increase enabled Sabena's A310s to extend their range by 900 miles (1,450 km), giving each jetliner a range of 4,260 miles (6,860 km) with a full payload.

In the 1984 summer timetable, the introduction of the Airbus A-310-222s on routes to West Africa was the main change. The A310s were given the following destinations: Dakar, Conakry, Kano, and Abidjan, all with two flights a week, and Monrovia and Douala with one flight each. From June 15 to September 15, the A310s were used on the Brussels-Kinshasa line only once a week for the summer period. For the rest of the week, that route was flown by other aircraft. The A310s were also scheduled with one weekly flight connecting Brussels to Athens and Beirut and three times per week to Tel Aviv.

In October, Sabena announced that it would remove the first class from its medium-haul network. Instead, the carrier would offer the less expensive business class on its 737-200s, which the airline believed was more suitable for the needs of its customers. Also, in mid-November, Sabena abolished its first-class services on some long-haul routes, which affected the DC-10s and Airbus A 310s. However, Sabena's 747s would keep their first-class services on the New York and Chicago routes.

Sabena's fleet in mid-1984 was as follows: 2 Boeing 747-129, 5 DC-10-30CF, 2 Airbus A310-222, 1 707-329C, 11 737-229, 4 737-229C. Light aircraft: 6 Cessna 310, 9 SF.260, 5 EMB 121 Xingu. There was one Airbus A310 on order.

1985

The year started with changes in Sabena's fleet. In January, the carrier announced that it was preparing to phase out its two 747-129s. Sabena's two aging Jumbos had logged more than 60,000 hours in the air, flying between 13-16 hours each day in 11 years since they were inducted into the fleet. They were believed to be two of the world's hardest-working airliners. Their replacement would be yet another Boeing Jumbo, the 747-300. Sabena stated that it had placed an order with Boeing for its first 747-300 Combi, with the extended upper deck. The new aircraft was going to be equipped with 16 seats in first class, 80 in business class, and 313 in economy, 409 seats in total. Business-class would occupy the whole upper deck, with the other two classes being on the main deck. Sabena would put the new 747-300s on its North Atlantic network, with flights to New York, Chicago, Detroit, and Atlanta.

Sabena's air traffic results for 1984, released in April, were outstanding. Sabena's load factor reached a record 73.8%, 5.4% higher than the previous year. In terms of passenger numbers, Sabena was first among all the European carriers in 1984, with a seat occupancy rate of 67%. The average for Europe that year was 65.4%. Concerning the strongest performing regions in 1984, East Asia led with 82.9%, while the North Atlantic network was close behind with 81.6%. However, the news was not so good on routes to the Middle East. The load factor in that region sunk from 51.6% to 20.6%. More than 2 million passengers flew with the Belgian carrier in 1984.

Sabena closed its balance sheet for the second year in a row with a positive result in its financials. That was remarkable for the Belgian airline, which from 1957 to 1982, constantly posted large deficits. In 1984, the company's receipts amounted to BFR 40.7 billion, while expenses reached BFR 38.8 billion, which resulted in a BFR 1.85 billion positive operating result for the airline. However, considering financial charges and stock dividends, the net profit carried forward was only BFR 26.4 million. However, that was better than being in the red again. Sabena's president, Mr. Van Rafelghem, said the results were relatively good. Still, Sabena was not yet in a strong enough financial position to be able to embark on a significant reinvestment program involving new airliners.

Photo 89. Sabena Boeing 747-329 'OO-SGD' taxiing t Brussels Airport. Delivered to Sabena on September 25, 1990. This was the last 747-300 ever built. It was withdrawn from use on August 28, 1999. (Photo from 'www.douglasdc8.com')

Photo 90. This is another photo of 747-329, 'OO-SGC,' on approach to London Gatwick in May 1987.
(Photo by Richard Vanderbord)

Photo 91. Sabena 747-329, 'OO-SGC,' about to land at London Gatwick in May 1987.
(Photo by Richard Vanderbord)

Photo 92. Sabena 747-329 'OO-SGC' was the fourth and final 747-329 delivered to Sabena. (Photo by Richard Vanderbord)

The airline did not sit on its hands in 1984. Sabena's business approach appeared to be better in line with market requirements. In 1984, the carrier refurbished the first-class cabins of its Boeing 747s to accommodate sleeper seats. While, on all DC-10 and 737 aircraft, it replaced 'First Class' with 'Business Class.' In terms of cargo, Sabena added some new services in that division. It provided a small parcel service for packaging up to 150 kg. The service also offered the sending documents to destinations where local postal services were sparse (read DHL, UPS, and FedEx). Another positive result was the creation, in March 1984, of a new subsidiary, the 'Belgian Fueling and Services Company' BFSC, with Sabena and Sobelair holding 68.5% of the stocks. It looked as if Sabena was starting to show promise. Could this new positive phase continue well into the future?

In the 1985/86 winter timetable, there were two additions. From October 29, Sabena started operating direct flights from Brussels to Kinshasa using its Boeing 747-229s. Although the flights were not a permanent fixture, they ran until March 29, 1986, at the end of the winter period. The new line to Kinshasa connected the two cities in 7 hours and 25 minutes. One flight flew non-stop on Saturdays, while a second flight on Tuesdays went via Athens. In addition, there were two other flights with Sabena's DC-10s to Zaire on Wednesdays and Sundays, with intermediate stops at Lagos and Libreville. In Europe, Sabena began flights to the industrial city of Gothenburg in southern Sweden and the Norwegian capital Oslo five times a week from the end of October. Oslo was added to Sabena's flight network again after a 35-year hiatus.

1986

The numbers for Sabena's air traffic performance for 1985 validated the Belgian airline's revival. Sabena increased its transport performance by 6.3% compared to 1984. In addition, Sabena's charter flights doubled during the year. Its aircraft transported 2,188,707 passengers in 1985, a 7.7% increase over the previous year. The favorable traffic trends were affected by several reasons. The Airbus A-310s, delivered in March and April 1984, created additional capacity for the Belgian airline, making it possible to expand its range of services on the North Atlantic network. Its newest US line, Boston, opened in April 1985. With the 1985/86 winter schedule, Sabena added Oslo and Gothenburg to its route network and used Boeing 747s on the Brussels-Kinshasa-Johannesburg route. The capacity offered by Sabena in 1985 was 1.53 billion ton-km, which was an increase of 8.1% over the previous year. As for air cargo, Sabena carried 108,696 tons in 1985, a rise of 11.6%. The carrier's number of flights also increased by 7.3% to 43,627. At the beginning of 1986, Sabena's fleet consisted of 2 Airbus A 310-222s, 5 DC-10-30CFs, 2 Boeing 747-229s, and 15 Boeing 737-229s.

In financial matters, Sabena made a profit, for the third year running, of BFR 1.2 billion and achieved a gross profit of BFR 783 million. After allocating BFR 437.5 million in dividends, Sabena was left with a positive balance of BFR 345.6 million, which would be given to a reserve intended for the renewal of Sabena's fleet. The improved result was due to implementing staff cutting measures and raising the company's equity to ease its debt burden. Moreover, Sabena sold and leased back two 747-100s during the year, freeing about BFR 2.46 billion in capital which went into a fund for a new 747-300, which was delivered in June 1986.

In April, a problem arose between Belgium and Zaire. The African nation cut all air traffic from Belgium and ordered its national airline, Air Zaire, to relocate its offices from Belgium to France. That decision followed a conflict between the Zairean airline and some European pilots who had challenged their dismissal from the Zairean airline. The pilots that Air Zaire laid off were part of a plan to replace foreign ex-pat pilots with cheaper African aircrews. However, the European ex-pat pilots claimed redundancy payments, which fell on deaf ears in Kinshasa. In addition, the foreign pilots of Air Zaire, primarily Belgians and French, constantly had difficulties with the Zairean airline, which had been behind on salary payments for several years. As a result, a Belgian judge decided to impound an Air Zaire DC-8 at Ostend soon after landing from Kinshasa in a tit-for-tat exchange. Following a case brought by Vincent Gauthier in 1984, the same sequence of events occurred. At that time, Zaire retaliated by seizing a Sabena DC-10-30 in Kinshasa, which was eventually released after Sabena paid BFR 7 million. At the same time, the Belgian government agreed to let Air Zaire's airliner leave the country. In a case of 'once bitten, twice shy,' Sabena avoided retaliatory action from Zaire in 1986 by canceling all of its flights to Kinshasa and flying directly to Johannesburg. Herman De Croo, Belgium's Minister of Transport, responded by denying that political concerns influenced the court's judgment of impounding the Zairean airliner. Zaire then accused Belgium of behaving disparagingly towards its former colony. Things eventually quietened down, so much so that Air Zaire and Sabena signed a new bilateral agreement for the Brussels-Kinshasa route in November. Both airlines would operate five weekly flights, with only two of these being operated in a pool. The other three would benefit only the serving airline (i.e., Air Zaire or Sabena). The agreement also covered the equal spreading of overheads that were pooled. Ultimately, the impounding of the Zairean airliner and the ensuing conflicts were minor problems.

In the summer schedule for 1986, Sabena announced that there would be two flights to Tokyo and three to Southeast Asia over the summer season, which would be operated by Sabena's DC-10s, whose cabins had been modified to improve passenger comfort. Regarding Africa, flights to Kinshasa increased from three to four per week on the Boeing 747s. Two of those flights continued their journey to Southern Africa. One flight was via Lagos, while the other flew to Brazzaville. In Europe, Sabena collaborated with Air France on the regional Antwerp-Paris route and with Lufthansa on the Antwerp-Frankfurt line. Those flights were operated by Fairchild aircraft of DAT, in which Sabena had recently acquired a significant stake. Finally, on the North Atlantic network, at the height of the summer season, 19 flights were provided to the following cities: New York, Boston, Detroit, Chicago, Atlanta, and Montreal. Twelve of those flights were on the 747s and seven on the DC-10s.

At the ordinary general meeting of shareholders on June 17, 1986, Mr. Carlos Van Rafelghem, chairman of Sabena, announced that the company's resources enabled the self-financing of the new Boeing 747-300, which had just entered into service with the carrier. The new 747-329 differed from its two 747-100 siblings by the more extended upper deck, enabling it to carry up to 385 passengers; 16 in first class, 68 in business, and 301 in economy. Each week, the 747-300 would operate six of the seven Brussels-New-York flights and one of the three Brussels-Montreal-Chicago flights. In addition, a second 747-329 was on order, with delivery scheduled for 1988.

The effort to rejuvenate Sabena's fleet was moving at a steady speed. In March 1987, a third Airbus A 310-329 was delivered to the carrier. Also, in August and September 1987, five new 737-329 jetliners, with accommodation for 124 passengers, were received by Sabena, including one for Sabena's subsidiary Sobelair.

Portugal featured in the 1986/87 winter schedule. The Portuguese city of Porto was now one of Sabena's stopovers, as was Costa Verde, a tourist and coastal region of northwest Portugal, which Sabena serviced every Monday via Lisbon, with its 737-229s. Spain was also a beneficiary of extra flights by Sabena. The Belgian carrier increased the number of its flights to Madrid from Brussels to 13 per week. Sabena also increased its flights to Rome from seven a week to nine. In addition, an extra flight to Gothenburg was added to the schedule in the Scandinavian region, making it six per week. In France, Bordeaux was linked to Brussels with five flights a week, which Sabena operated in a pool with Air France. In November, Sabena announced that it would replace its DC-10s on routes to East Africa with its new Airbus A310-300 once the aircraft got delivered in February 1987. Sabena was already using 2 Airbus A310-222s on its short and medium-high-density routes.

1987

In February 1987, Sabena and the Scandinavian airline SAS announced that they held talks on entering into a close collaboration. However, a merger of the two companies was ruled out, but Sabena and SAS contemplated setting up a new company that would be given a new name. However, there were a few reservations that had to be ironed out. The CEO of SAS, Jan Carlzon, discussed the merger plans during a meeting with representatives of the unions, which led to great unrest among SAS employees in Denmark, Sweden, and Norway due to fears of the relocation of SAS's international scheduled services to Brussels. For his part, Sabena chairman Carlos van Rafelghem stated that he was opposed to a complete merger with SAS. He did say that it might be possible for Sabena to become the fourth partner in the Sweden, Norway, and Denmark triad. However, Sabena's presumed loss of identity must first be considered if the Belgian carrier was to join SAS in forming the new airline. Many observers in Belgium were worried that the name Sabena would disappear for good if the deal went through.

Just four months later, Van Rafelghem stated that Sabena and SAS would not merge. Still, Sabena and SAS discussed forming an international subsidiary airline to handle some of their current independent activities. Van Rafelghem's declaration responded to suspicions that the Belgian airline would join the SAS trio as a fourth partner. He did say that there would likely be increased cooperation, but it would not extend to domestic flights, which would remain the responsibility of the various national carriers. In essence, the negotiations had cooled off by June because the Belgian side feared that a merger with SAS would sooner or later deprive Sabena of its identity. The so-called collaboration between Sabena and SAS never took off.

There were some new destinations in the 1987 summer timetable, and they were all in Africa. Sabena added Niamey (Niger), Lome (Togo), Cotonou (Benin), and Luanda (Angola) to its route network. With the addition of the four new cities, Sabena provided flights to 27 destinations in Africa. The Belgian carrier had provided flights between Europe and Africa for more than 50 years. The latest additions to the network were scheduled with the following flights starting from March 29, Sabena's Airbus A-310-300s would fly once a week to Niamey, while the 747-300s would fly once a week to Luanda via Lagos and Kinshasa. In addition, Lome and Cotonou would be provided with two flights each per week. Previously, on March 6, Sabena put the new Airbus 310-300 into service on some of its other African and European routes. The Airbus flew from Brussels to Dakar, Monrovia, Conakry, Abidjan, and London. In addition, the airliner was also put into service on a few flights to Paris. Meanwhile, on the North Atlantic network, Sabena opened a new service to Toronto in the spring of 1987, which its DC-10s would service.

In its air traffic performance for 1986, Sabena recorded traffic results roughly comparable to 1985. The Belgian company was satisfied with the results despite the difficult economic conditions worldwide. Sabena's seat occupancy rate was 65.9% throughout its network compared to 65.7% in 1985. In 1986, the Belgian carrier transported 2,233,220 passengers, flew 32,814,610 miles (52,809,997 km), and completed a total of 87,608 flight hours. Sabena's transatlantic traffic suffered from the lack of interest by American passengers to fly to Europe in 1986 due to terrorist acts in certain European countries, the Chornobyl disaster, and the declining US dollar. Sabena was not the only European airline with a lack of American passengers. All other European airlines faced the same problem in 1986. It was just a bad year for the flights on the North Atlantic network. Sabena did increase its passenger traffic in the other regions. In Europe, traffic was up by 4.4%, while it increased by 2.5% in Africa. Over in Asia, traffic rose by 4.3%. On the flip side, traffic dropped in the Middle East by 2.1%. However, the North American decline was the worst, as traffic in the US and Canada dropped by 9.7%. Financially, the Belgian carrier saw its profit fall from BFR 345.5 million in 1985 to BFR 146.8 a year later. However, in 1986 Sabena did make another profit, the airline's fourth in a row, but it couldn't match 1985, which was the Belgian company's best financial year in 28 years.

In November, the talks between Sabena and SAS on achieving an agreement on close collaboration were faltering. Sabena then started negotiations with Swissair on a possible partnership before 1992. Sabena was looking for a partner in Europe due to a common internal market that would emerge within the European Union (EU). That would bring more competition between airlines and the probability of lower salaries for the staff. What was going to be unleashed upon the EU in 1992 was basically 'Deregulation v2.0', after the American model, which sunk some of the most iconic US airlines like Pan American Airways, TWA, Eastern Air Lines, Braniff International Airways, Northwest Airlines, and so on. The European Union's transport ministers agreed to 'liberalize' Europe's skies. That was a euphemism for deregulation. This action would have far-reaching implications for Europe's airlines. The European ministers approved a package of measures in December 1987 that would give airlines more flexibility in setting tariffs, deciding flight frequencies, and other comparable activities.

Back to the Sabena and Swissair collaboration talks, in November, a Sabena official stated that the discussions with Swissair were focused on cooperation on common routes and in the operational area, but not a complete merger. At the same time that Sabena started talks with Swissair, it was in discussions with Austrian Airlines and KLM.

In the ongoing long-range fleet modernization program, Sabena ordered five Airbus A340-300 Combis, officially becoming the tenth airline to order the type. The aircraft would be used on routes operated by Sabena's McDonnell Douglas DC-10s. The A340-300s were scheduled to be delivered in 1993 and 1994. The four-engine jetliners offered accommodation for 200 passengers in all three classes, first, business, and economy class. They could also carry 29 tons of freight. The Sabena Combi versions had a side door at the rear of the fuselage to permit the loading of four freight pallets onto the main deck. The A340-300 flew for the first time on October 25, 1991, just 21 days after rollout. It was powered by four CFM International CFM56-5 engines. The airliner had a range of 8,390 miles (13,500 km) and a maximum speed of 568 mph (914 kph). In 1987 Sabena's fleet was as follows: 1 747-329 Combi, 2 747-129Combi, 5 DC-10-30CF, 1 Airbus A310-329, 2 A310-222, 4 737-329, 11 737-229, 4 737-229C. Light aircraft: 9 SF.260, 5 EMB 121 Xingu. There were also the five A-340-300s on order.

1988

There was some hope for Sabena. KLM and the Belgian carrier reached an agreement in February 1988 on long-term cooperation. The two airlines would liaise in joint air operations. Sabena discarded the offer to join SAS because it would be swallowed up by the Scandinavian airline. KLM was not looking for a merger when the agreement was signed with Sabena. The Belgian airline was adamant that it was not looking for a union when it signed the contract with KLM. A Sabena official stated that the cooperation with KLM should be limited to air transport operations only. Sabena was not seeking collaboration in the field of maintenance or catering. KLM was okay with that. The Dutch airline was much larger than Sabena, two and a half times Sabena's size.

Sabena still had unpleasant memories of the events of 1975, when there were discussions of merging all three airlines from the low countries, Sabena KLM and Luxair, and forming a super Benelux airline. Merger talks broke down mainly because the French-speaking population of Belgium wanted nothing to do with such plans. The Belgians feared, not without justification that many staff cuts, route network culling, and other belt-tightening measures would have had to take place by Sabena as a result of the merger. Moreover, KLM would have been the kingpin in the new Benelux airline, which did not go down well with Sabena. In a strange twist of fate, the cuts that Sabena would have had to make in 1975 if it had merged with KLM and Luxair were eventually made in the early 1980s, when the Belgian airline was forced to reduce its workforce from 20000 to 9,000. It also suspended the destinations with low load factors and ceased buying new airliners like there was no tomorrow. After a few years of being in the black, Sabena was prepared to take on new ventures.

In April, Sabena's Chairman, Van Rafelghem, stated that he had spoken with other European airlines, Air France, Lufthansa, Swissair, and Luxair. He claimed that Sabena's aim was for a formula that maximized the benefits of the Brussels Airport hub. According to Van Rafelghem, any future cooperation between Sabena and other airlines would be limited to a jointly-owned subsidiary under European law. The project partners would need to form a new jointly-owned airline to coordinate all operations. In mid-June, there were tensions in Sabena's search for partners. The Walloon (French Belgian) Socialist Party, Guy Spitaels, expressed reservations about Sabena's cooperation with only KLM. The old festering wounds were appearing again. The Belgians were wary of any affiliation with KLM. Then, in November, there were rumors that the French-speaking government parties in Belgium wanted to keep KLM from becoming a Sabena partner. This information was leaked in a private discussion paper circulated at the Walloon Socialist Party's headquarters. The leaked memorandum confirmed the distaste many French-speaking Belgians felt against any merger between KLM and Sabena. Previously, the Walloon government parties had already stipulated that the cabinet be given a veto right over Sabena's choice of partner. The memorandum pointed out the dangers of a Dutchification of Sabena by KLM. The Walloon Socialist Party stated, "No Walloon will allow such a significant national object (Sabena) to fall prey to a Dutch-Flemish seizure of power." However, no economic or financial rationale was presented in opposition to a collaboration with KLM by the Socialist party. According to the Party, a merger between Sabena and Air France would be a more realistic scenario for the Belgian airline because the French-speaking Belgium population contributed more to Sabena than the Belgian Dutch. The road to finding a partner for Sabena was not going to be an easy one.

In February, Sabena announced that it had joined the European Galileo computer reservation system consortium. The number of partners using the system thus increased to 10, including Alitalia, Aer Lingus, Austrian Airlines, British Airways, KLM, Swissair, TAP Air Portugal, and Covia. Sabena invested BFR 300 million in the project. The technology was developed in 1971 by a Chicago-based corporation called Covia, a United Airlines subsidiary. The Galileo computer reservation system, created in July 1987, allowed users to access a wealth of information on flight schedules and ticket costs, hiring a car, and reserving a hotel room. As a result, Galileo became one of the world's largest airline reservation systems, with a 26.4% share of worldwide computer reservation bookings by the early 2000s. Not to be outdone, four other major European airlines, Air France, Lufthansa, Iberia, and SAS, formed a joint company to set up and operate their European reservation system called 'Amadeus.'

In April, Sabena reported that its 1987 net profit fell by 52% to BFR 70.4 million compared to the 1986 figure of BFR 146.8 million. Sabena was quick to name the culprits for the negative operating result for 1988, strikes, foreign exchange losses, and the disruption of services in Zaire. The results showed a decline of 5.4% in cargo transport, which was partially offset by the excellent performance of passenger traffic, which rose by 7.4%. Moreover, Sabena's productivity was up by 1% when combining passenger, cargo, and mail traffic.

In 1988, Sabena went on an airline acquisition spree. In April, the Belgian airline exercised an option for a second Boeing 747-300, which would replace one of the fleet's older 747-129s. Later, in June, the airline placed a significant order for 16 Boeing 737s, 4 of which were 737-400s and 12 of which were 737-500s, all powered by General Electric/Snecma CFM56 engines. The 737-300s, which were new additions to the fleet, could accommodate 126 passengers. On the other hand, Sabena would be able to provide 144 seats for scheduled flights and 168 seats for charters with the stretched 737-400. Two of the 737-400s were ordered by Sabena's subsidiary, Sobelair. The 737-500, on the other hand, was substantially smaller than the 737-400, having only 111 seats. Despite this, it had a somewhat bigger capacity than the 737-200 it would replace. The 737-400s and 500s were scheduled to be delivered in 1991 and 1992. With the large order in hand, Sabena and Sobelair's fleets would grow to 23 737 airliners by 1992, ready for the newly liberalized European skies.

Sabena's fleet in 1988 consisted of 1 747-329 Combi, 2 747-129 Combi, 5 DC-10-30CF, 1 Airbus A310-322, 2 A310-222, 4 737-329, 11 737-229, 4 737-229C. Light aircraft: 9 SF.260, 5 EMB 121 Xingu. Sabena's order book was full, it had 5 A340-300, 1 747-300, 4 737-400, and 12 737-500 on order.

1989

Sabena started the year on the wrong foot, as another feud blew up with Zaire. In January, the Belgian government started the dispute when it strongly protested Zaire's decision to cancel the friendship treaty and the treaty on development cooperation between the two nations. Zaire was in no mood to compromise and suspended the repayment of the Zairean debt to Belgium of about BFR 43 billion. To make matters worse, the Zairean Government cut in half Sabena's flights to Zaire and even threatened to withdraw Sabena's landing rights to the country permanently. If that were to happen, it would result in a financial disaster for Sabena, as the Belgian carrier received around 40% of its income from flights to Africa. Even worse, if the dispute led to a complete split between the two countries, 12,000 people would lose their jobs in Belgium. Furthermore, 3,000 Belgian development workers in Zaire also ran the risk of being expelled. The Government in Brussels called the reduction of the number of weekly flights of Sabena from four to two a flagrant violation of the existing aviation agreement between Belgium and Zaire. Finally, sanity prevailed in late January when Belgium and Zaire agreed to hold a ministerial conference shortly to improve relations that had fallen apart. Zairean President Mobutu said he wanted to settle the conflict with Belgium. However, nothing came of the meetings in the ensuing weeks. It took about seven months for the acrimonious atmosphere between the two nations to ease off. The dispute eventually ended in mid-July, when an agreement was concluded between Belgium and Zaire that reduced Zaire's debts to Belgium by approximately BFR 18.6 billion. Moreover, Zaire was allowed to pay off part of the remaining debt with its national currency. In return, Sabena was once again permitted to fly unimpeded to Zaire.

Cooperation and partnership were the two buzzwords for Sabena in 1989. The Belgian airline was hunting for partners in its quest to feel secure before the advent of Europe's 'liberated' skies in 1992. In February, a cooperation agreement was reached between Sabena and Swissair. In terms of the deal, Sabena would suspend its routes to Jeddah (Saudi Arabia) and Dar es Salaam (Tanzania) on March 26, 1989. Swissair would pick up the slack from that date on. In addition, Sabena would provide a feeder service from Brussels to Zurich and back every Monday and Thursday. Swissair would take the Belgian passengers from Zurich and fly them to Saudi Arabia or Tanzania. To make their cooperation visible to all, the two airlines designated these flights as the Swissair-Sabena joint service with a standard flight designation code: SR/SN.

Photo 93. Sabena 737-329, 'OO-SYA' on the tarmac at Salzburg Airport in Austria on September 29, 1990. Delivered to Sabena on April 28, 1989. It was in the fleet until Sabena's collapse in 2001. (Photo by Michael Bernhard)

Photo 94. 737-329 'OO-SDX' at Brussels Airport. Delivered to Sabena on September 17, 1987. It was taken out of operation on the final day of Sabena, November 7, 2001. (Photo from 'www.douglasdc8.com')

Photo 95. 737-329, 'OO-SDV,' taxiing at Brussels Airport. Delivered to Sabena on August 25, 1987. It flew for Sabena until the last day for the Belgian national flag carrier, November 7, 2001. (Photo from 'www.douglasdc8.com')

Photo 96. Sabena 737-329, 'OO-SDW' in Sabena's final livery at Heathrow Airport in May 1996. Delivered to Sabena on August 26, 1987. Like the other 737-300s, this 737 also lasted until November 7, 2001. (Photo by Richard Vandervord.

Sabena World Airways - Part 1

The partnership that Sabena had been looking for the previous three years to better cope with increased competition in Europe's skies from 1992 had finally arrived for the Belgian company. In April, while discussions were ongoing, Sabena's head, Carlos Van Rafelghem, announced the creation of a joint subsidiary with the Dutch national airline, KLM, and the UK national carrier, British Airways (BA). The new airline was named Sabena World Airlines (SWA) and would become operational on January 1, 1990. Unfortunately, Mr. Van Rafelghem announced the news a bit prematurely. That left KLM and BA a bit jolted. When asked by the press for their views, British Airways declined to comment, and KLM responded by stating that everything reported was premature. An official said, "We've been discussing our intentions with several possible partners. We will only provide information when we have concrete news to share." In essence, KLM and BA were right. There were a few more hoops to jump through before confirmation could be made. The agreement had yet to be approved by the directors of KLM, BA, and the European Commission because the Belgian Government held a 53% stake in Sabena. Finally, in early July, it was confirmed that Sabena would have 60% of the capital of its subsidiary SWA, with British Airways and KLM each taking 20%. The holdings by the trio would enable an expansion plan to be implemented as quickly as possible so that the development of Brussels National Airport as a major hub for the intra-European and international air traffic could be completed on time. Once the Brussels Hub became operational, BA would reschedule its African flights via Brussels and use the Belgian capital as a continental hub. In addition, Sabena would have fifth freedom rights at Manchester airport, which was to be expanded as a base for the SWA group's joint transatlantic operations, which would involve frequent shuttle flights between Manchester and the two cities of Brussels and Amsterdam.

SWA would eventually become a holding company with 16 independent subsidiaries, and it would take over Sabena's air operations on January 1, 1990. Sabena, British Airways, and KLM finally signed an agreement on December 13, 1989, after months of arduous talks in Brussels. SWA was jointly capitalized by all three airlines. The three airlines then informed their respective governments and the European Commission (EC) that they wanted to seek approval for the deal. Carlos van Rafelghem was selected to be the chairman of SWA and the president of Sabena's holding company. The participation of British Airways and KLM would have no bearing on the autonomy of the three carriers. They would still be rivals in the future. SWA would operate on both the European and intercontinental markets, utilizing Sabena's global portfolio of landing rights.

The SWA group was on the European Commission's radar. In May, British Midland Airways (BMA) chairman Michael Bishop issued a warning, claiming that British Airways and Sabena's influence in five hubs within 250 miles of London was anti-competitive. The addition of Amsterdam to the earlier list of Manchester, Birmingham, Heathrow, Gatwick, and Brussels and the inclusion of KLM in the arrangement had heightened his anxieties. The EC stated that it would remain neutral in the matter until an agreement was revealed and a comprehensive complaint was filed. A KLM executive said, "We understand that BMA fears a monopoly position, but we believe that, notwithstanding our cooperation with other airlines, competition will be considerably fiercer in 1993 (after the beginning of deregulation)." BMA waited for more details of the deal to be released before finalizing its complaint to the EC.

All was ready for the new airline to start operations on January 1, 1990. Sabena's management was ecstatic that the great future they had hoped for had finally arrived. What could go wrong?

In January, Sabena ordered a second Boeing 747-300 Combi. The two aircraft would eventually replace the carrier's two 747-129s, which were operated under a lease-back deal. In April, a second order was placed by Sabena. This time the airline agreed to place an order for four Airbus A340 long-range airliners and took an option on a further five. Sabena expected to receive the A340s in 1992. Sabena's third order was for two

British Aerospace 146-200s for its Delta Air Transport (DAT) subsidiary. The aircraft would be used on regional routes from Antwerp and Brussels. The Bae 146s would replace DAT's Fokker F.28 twinjets. Delta Air Transport took delivery of its first of eight BAe 146-200s on December 1, 1989. The 146-200s could accommodate up to 100 passengers. They were powered by four Lycoming ALF 502R-5, Avro RJ: Honeywell LF 507-1F engines, each with 6,990 lbf of thrust (31.1 kN). They had a maximum speed of 490 mph (789 kph) and a range of 2,268 miles (3,650 km).

In December, Sabena was forced to take over the training of its pilots from the Belgian government from January 1, 1990, following the cancellation of the government's funded pilot training scheme. Due to economic considerations, the government cut the BFR 94 million subsidies. As a result, Sabena, which needed 250 pilots between 1989 and 1994, established its own Belgian Aviation School (BAS) by utilizing the government-owned fleet of SF.260s and EMB 121 Xingus aircraft and the Xingu simulator. BAS relocated its principal training to the United States to avoid Belgium's inclement weather. However, advanced pilot training to an air transport license level continued at Brussels Zaventem airport.

Regarding Sabena's air traffic performance in 1988, the numbers were satisfactory. The Belgian airline carried around 260,000 more passengers in 1988 than in 1987, increasing by 10.3%, while cargo tonnage rose by 17.9%. These two statistics provided insight into Sabena's air traffic growth, which was notably visible on the Asian network, thanks to the third flight to Tokyo.

Sabena added 13 more European stopovers for the summer season schedules, including two capitals, Bern and Ankara. Three flights per week were provided for Bilbao and Naples, four for Glasgow, Edinburgh, Hanover, and Bordeaux, and five for Basel and Bern. Newcastle, only recently included in Sabena's route network, benefited from 11 weekly flights. In Italy, Venice was scheduled with six flights per week. There were 59 weekly European flights made available for the summer of 1989. Sabena was also able to introduce two new flights to Barcelona, four to Porto, and six to Gothenburg and Oslo, thanks to the arrival of two extra Boeing 737-300s, which joined the four already in service. In addition, Sabena's collaboration with English airlines improved services to Manchester and Birmingham. Sabena also added flights to its long-haul network. The Belgian carrier offered flights to Chicago, Bombay, and 11 African cities. In addition, the Belgian airline operated four flights to Kinshasa.

Sabena's fleet, at the end of the 1980s, consisted of the following aircraft: 1 747-329 Combi, 2 747-129 Combi, 5 DC-10-30CF, 1 Airbus A310-322, 2 A310-222, 6 737-329,10 737-229, 3 Fokker F.28, and 5 EMB-120 Brasilia. The aircraft on order were, 5 A340-300, 2 737-329, 2 737-429, and 12 737-529.

Delta Air Transport – DAT

Delta Air Transport (DAT) was founded at Antwerp Airport in 1966 by Frans Van den Bergh, a real estate broker, to provide passenger and cargo charters and air taxi services around Belgium from Deurne airport initially and later from Zaventem airport. Douglas DC-3s were used to launch commercial flights, with the first route being Antwerp-Deurne. Delta opened the second route Antwerp-Amsterdam with Convair 440 Metropolitans. In 1971, Delta Air Transport received its first DC-6, followed by two more in 1972. An ex-United Airlines Boeing 720-022 was leased in May 1974, allowing DAT to offer worldwide charter flights, which proved unsuccessful.

Consequently, the airline was bankrupt by the beginning of 1975. In that year, Sabena became involved with the airline as a shareholder. The Belgian national airline purchased 33.4% of DAT's shares.

In 1976 Tony Vangrieken, a flight crew inspector for KLM and a training and operations instructor from 1969 until 1976, joined DAT as general manager on behalf of KLM and saved DAT from its new owners who wanted it liquidated. Tony Vangrieken eventually became DAT's main shareholder.

In 1986, Sabena became the majority shareholder when it purchased 79% of the shares. With the involvement of Sabena, DAT's fleet expanded as the Belgian national carrier put in an order for two Embraer EMB-120 Brasilia turboprops. Then, in 1987, a Fokker F28-3000 was purchased, and flights to London and Zurich began. Finally, in mid-1987, orders for ten more Embraer EMB-120 Brasilia aircraft were placed, of which two were delivered in March and April of 1988. The EMB-120 Brasilia aircraft were placed on the following routes from Brussels: Lyon, Toulouse, Bordeaux, and Marseilles.

In 1990, there were significant changes in DAT's fleet. All the older Fairchild FH-227 turboprops were retired. The new additions to the fleet were one Fokker F28-4000, one EMB-120 Brasilia, and four British Aerospace BAe-146-200 jetliners. In 1991, Vangrieken was appointed chairman and managing director, and he added two more BAe-146-200s and eight EMB-120 Brasilia aircraft to the fleet. By 1991, DAT's route network comprised the following destinations: (UK) Manchester, Newcastle, Leeds, Bristol, Edinburgh, and Glasgow. (Germany) Munich, Dusseldorf, Hanover, and Stuttgart. (Italy) Florence, Naples, Turin, and Venice. (France) Bordeaux, Lyon, and Marseilles. (Portugal) Porto. (Spain) Bilbao. (Scandinavia and Finland) Stockholm, Oslo, and Helsinki). (Czech Republic), Prague and Luxembourg. After the bankruptcy of Sabena in 2001, its profitable subsidiary DAT was acquired in 2002 by SN Air holding. Consequently, DAT was incorporated into Brussels Airlines.

Photo 97. A Delta Air Transport FH-227B, 'OO-DTA,' speeding down the runway at Amsterdam Airport Schiphol. Delivered to DAT in May 1977. It was scrapped in December 1994. In 1997 it ended up in Uruguay, in a deteriorated state. (Photo from 'www.douglasdc8.com')

Photo 98. Another image of FH-227B, 'OO-DTA' in a mixed Sabena/DTA livery. (Photo by Eduard Marmet.

Photo 99. A Delta Air Transport DC-6B, 'OO-FVG,' at Munich-Riem Airport (Munich) on February 1, 1972. It was delivered to DAT in December 1971. In August 1973, it was re-registered 'OO-VGF.' It was finally sold to the Canadian company specializing in retrofitting firefighting aircraft, Conair Aviation, in January 1975. (Photo by Norbert Kröpfl)

Photo 100. Another FH-227B 'OO-DTB' in a mixed Sabena/DAT livery. Delivered to DAT on April 25, 1977. It also ended up in Uruguay and was broken up in January 2001. (Photo by Udo K. Haafke)

Photo 101. FH-227B 'OO-DTB' is seen here at Jersey Airport in May 1980. (Photo by Richard Vandervord)

The Turbulent Decade
1990-1999

1990

Sabena World Airways - Part 2

Following the agreement on the foundation of SWA, Belgium had enormous plans. The Belgian government expected Brussels to become Europe's most significant hub. However, Zaventem Airport, which was quite old and worn, would need to be repaired entirely at hundreds of millions of Belgian francs in the coming years. Furthermore, Sabena's European fleet would be significantly expanded, with at least 50 aircraft. A total investment of approximately BFR 92 billion was planned to purchase new airliners. However, it was still unclear who would fund Sabena's fleet growth since the carrier was mainly Government-owned.

In May, a Belgian parliamentary committee's adoption of a new law which ratified Sabena World Airline's monopoly of scheduled flights from Brussels Airport drew heavy criticism from other airlines. The Belgian charter airline, Trans European Airways (TEA), spoke up against SWA's dominant position at Brussels Airport. TEA also wanted to operate its own international scheduled services from Brussels. TEA's chairman, Georges Gutelman, expressed confidence that the European Commission (EC) would recognize the SWA situation and take appropriate action. The EC could not approve the creation and misuse of a dominant position at Brussels Airport by three major Community airlines, abetted by the Belgian government. British Midland Airway (BMA) was in support of TEA and also lodged complaints with the EC about what they saw as an anti-competitive practice by SWA. In early July, the EC stated that it looked into the situation with SWA and, in principle, had no objections to the union of Sabena, KLM, and British Airways. However, the European body wanted to examine the unique project from a competitive point of view and, if needed, implement some changes. Sir Leon Brittan, European Commissioner for Competition, announced that the contracting parties had been invited to negotiate with the EC. They could look into specific contractual terms between the three airlines that could make it difficult for other European Community competitors to compete in the market. Then in July, the Brussels Commercial Court ruled that Sabena World Airways did not have the exclusive right to monopolize international flights from Brussels Airport for its parent company Sabena. The Belgian carrier promptly summoned its lawyers to file an appeal against the decision. Sir Leon Brittan was deliberating whether SWA's monopoly violated EEC competition regulations. However, his situation was aggravated by EC Transport Commissioner Karel Van Miert's support for monopoly positions for national flag carriers.

The situation was getting complicated for SWA. In September, alarm bells rang when it was announced that Sabena made a significant loss of around BFR 1.3 billion in 1989. Moreover, the carrier was expected to make another loss in 1990 of approximately BFR 3.7 billion. With the enormous losses incurred by Sabena, the airline required a BFR 6 billion financial injection to avoid financial ruin. Therefore, new shares were issued to be sold to private investors. As a result, the Belgian Government announced it would reduce its stake in Sabena from 54% to 26%. Semi-government entities held the remaining Sabena shares. The partnership between Sabena, KLM, and British Airways was unaffected by the partial privatization. KLM and BA each held a 20% share in SWA, with Sabena owning 60%. Furthermore, both KLM and BA had invested more than BFR 1.8 billion in SWA.

In November, the Belgian transport minister, Jean Luc Dehaene, announced that Pierre Godfroid would be the new CEO of Sabena, beginning on January 1, 1991. His priority was to develop a refinancing strategy that included obtaining new risk-bearing capital. The plan would be subject to the condition that the shares of the Belgian state would not be diluted past 26% of the total outstanding share capital and that the Belgian Government, together with the institutions controlled by it, held a 51% majority in Sabena. However, Dehaene refused to specify whether Sabena's financial difficulties with Sabena World Airlines would impact the alliance with KLM and British Airways.

Amid legal and political battles, SWA was focused on planning to serve some 75 cities with 400 daily flights and providing rapid connections from Brussels National Airport. Thanks to the novel plan, new routes would be opened by 1990 to Warsaw, Budapest, Prague, Salzburg, Malta, Venice, Bristol, and Seville. In addition, SWA stated it would provide three daily flights to the following cities in its European route network: Cologne, Strasbourg, Norwich, Bristol, Newcastle, and London to Gatwick and Stansted airports. The second group of cities would be provided with two daily flights: Venice, Helsinki, Hamburg, Seville, Pisa, Bologna, and Valencia. SWA's President, Carlos van Rafelghem, predicted that with the different European destinations and flights, SWA would triple its European air traffic by the end of 1995. SWA was also planning to strengthen its long-haul services considerably. SWA prepared for an intensification of flights to New York on its North Atlantic route network, with a second daily connection by Airbus A310-300s, and to the cities of Chicago, Montreal, and Toronto. SWA was also planning to open two new routes to the US, Brussels-Manchester-Los Angeles and Brussels-Gatwick-Houston. In the African and East Asian regions, Sabena opened a new line to Banjul, Gambia, and offered a third weekly flight to Bombay, Bangkok, and Singapore. Carlos van Rafelghem estimated that SWA's first-year turnover would be about BFR 87 billion. However, according to him, the projected expansion of Brussels Zaventem Airport was critical to the success of SWA's objectives.

In March, a dispute occurred between Delta Air Transport (DAT), Sabena, and the Belgian Pilots' Association. DAT applied to Belgian authorities to be given the right to fly several of its routes under its banner rather than Sabena's. The decision was made in response to complaints by the Belgian pilots' organization, ABPNL, that Sabena had been exploiting DAT as a source of cheap labor since the introduction of the British Aerospace 146 as part of the expansion program. The dispute triggered the grounding of DAT's two BAe 146s. DAT also lost its routes from Brussels to Stuttgart, Florence, Oslo, and Stockholm. The argument was centered on BAe146 crew pay. DAT crews agreed to operate Fokker F.28s and smaller planes for less money than Sabena pilots. The ABPNL, on the other hand, wanted Sabena to pay DAT's pilots Boeing 737 crew rates, which were higher, on the four-engine BAe 146 airliners. The union also demanded that Sabena, DAT, and Sobelair pilots share a seniority list. Sabena's president, Carlos van Rafelghem, agreed to withdraw the BAe146 from service to save a comprehensive agreement on a 10% salary increase for Sabena 737 pilots in exchange for a 4,125-hour higher annual productivity agreement. DAT would continue to fly under its flight numbers, but the BAe 146s were unlikely to be used by the airline again. Carlos van Rafelghem directed DAT to dispose of the two BAe 146s in any way feasible. A new regional airline named DAT II was formed and would operate under SWA's umbrella. The airline was planned to be located in either the Netherlands or the United Kingdom. Due to the Bae 146s being from the UK, they were given UK registration and flown by British pilots. DAT intended to acquire eight BAe146s, with two of the four on firm order having been in service since 1989. According to BAe, the other two were delivered in June and July 1990.

Then in June, there was a breakthrough. It was agreed that until April 1, 1991, the Bae 146s would be flown by British pilots provided by the manufacturer, Bae. Then, beginning April 1, 1991, Sabena pilots would replace the British pilots under the agreement between Sabena and the unions. DAT pilots would be recognized as 737 pilots and receive higher salaries. The union demands that DAT pilots be included in Sabena's seniority list and that DAT pilots operating BAe 146s on Sabena routes should be paid Sabena salaries, was approved. The ABPNL acknowledged that DAT was and would continue to be a low-cost airline. However, its wages would generally remain lower than Sabena's, which applied only to Fokker F.28 and Embraer Brasilia aircraft, not the Bae 146s. The union also agreed to Sabena's proposal for increased productivity. The number of annual flying hours per pilot was raised from 625 to 725, and time off between flights was cut from 16 hours to 10 hours, while the daily number of sectors increased from 4 to 6. Furthermore, both DAT's Bae 146s, serving on Sabena routes, were repainted in the livery of SWA.

In June, Sabena released its financial report. Unfortunately, the results were not what management expected. Sabena experienced an operating loss of about BFR 1.3 billion in 1989, following a profit of BFR 1.2 billion in 1988. Despite this, Sabena posted a positive net result due to extraordinary income. Zaire's strikes and service disruptions were blamed for the dismal operational result during 1989. Furthermore, a BFR 1.5 billion loss was predicted for 1990, owing primarily to the costs of developing Brussels Airport Zaventem into a European hub. In December, Sabena's aviation operations were transferred to the new Sabena World Airlines holding company.

Also, in June, a Sabena A310-322 became the first civil aircraft to use satellite communications for air traffic control during planned operations, which was fantastic news for the Belgian flag carrier. The Sabena A310-322 delivered positional reports and navigation data from its flight management system via an Inmarsat satellite and European Space Agency ground station in Madrid to Eurocontrol and the UK Civil Aviation Authority. As for Sabena's fleet, in early 1990, it had the following aircraft on its list: 1 Airbus A310-322, 2 A310-222, 1 Boeing 747-329 Combi, 2 747-129 Combi, 6 737-329, 11 737-229, 5 McDonnell Douglas DC-10-30CF. The following aircraft were on order: 10 A340-300, 1 737-329, 2 737-429, 12 737-529.

1991

After the failure of merger talks with SAS, Sabena experienced yet another broken dream in its quest to find partners. In January, British Airways and KLM backed out of the SWA contract. The reasons for doing so were Sabena's financial situation which needed reorganization to control its rising losses. In 1990, Sabena experienced a significant financial setback due to enormous losses, creating major internal unrest for the carrier. The losses ran into the billions of Belgian francs. There was also the concern that the European Commission would not allow the SWA alliance to continue due to unfair competition. Sabena World Airline's projected creation of Brussels hub services was also an issue. They had been delayed for many months due to new Belgian Government laws enacted in 1991, which reduced Sabena's privileges and Brussels Airport slots in response to monopoly charges leveled by other airlines such as TEA. The changes impacted Sabena World Airline's prospective domination of the hub. KLM was not sorry about the termination of its collaboration with SWA. However, it sought new types of cooperation with the Belgian carrier, particularly given the importance of Brussels Airport due to its location and the Belgian carrier's scheduled services to Africa, which piqued the Dutch airline's interest. Thus, Sabena saw the end of its grand ambition to be a major airline in a central location in western Europe.

The losses in 1990 were terrible, which forced Sabena to undergo drastic restructuring. Self-preservation kicked in after the stepping down of Managing Director Carlos van Rafelghem on

December 31, 1990. The newly appointed Sabena MD, Pierre Godfroid, delivered his plan for the Belgian carrier's reorganization, labeled; 'The Sabena 2000 plan'. Godfroid stated that Sabena's subsidiaries, Sabena World Airlines, Sabena Technics, and Sabena Catering, would be merged into Sabena on February 1, 1991. Furthermore, Godfroid said the carrier would cut services to the following North American and East Asian routes: Bombay, Bangkok, Chicago, Atlanta, and Montreal. As for New York, only one airliner would be deployed to the Big Apple instead of two. The worst consequence of the reorganization was that Sabena had to lay off 2,200 employees, which was one-fifth of the airline's workforce. Half of the staff were forced to resign with a golden handshake and social guidance. The other employees were discharged over time. In February, about 800 employees had already taken early retirement.

As for the losses in 1990, Sabena was hit by a financial tsunami. The company reported a net loss of BFR 7.152 billion in 1989. The worst part about it was its subsidiary, Sabena World Airlines was the guilty one, as it posted a loss of BFR 7.5 billion. Sabena's losses amounted to BFR 21 billion in the previous three years. On the other hand, Sabena recorded a positive operational profit of BFR 169.4 million in April 1991, helping the airline exit the red for the first time in 18 months. The Belgian government pledged a total of BFR 35 billion in aid to Sabena to save it from bankruptcy. Breaking down the figures, the Government would inject BFR 10 billion into the capital of Sabena. It would grant the airline an additional BFR 9 billion as soon as it found a partner. Moreover, the state would cancel the BFR 16 billion in debts owed by Sabena to the State. The European Commission usually would have taken a hard line with subsidies of that magnitude; however, the EC dropped its objections in Sabena's case. There were conditions, of course, Sabena would no longer be able to receive Belgian Government support, and the company would no longer receive any form of preferential treatment from the Belgian government. Sabena was still mainly in the hands of the Belgian state, and the aid was considered anti-competitive. Sabena had always tried to keep other airlines out of the market. The earlier case brought against Sabena by its charter competitor Trans European Airlines (TEA) was lost by the Belgian flag carrier at the European Commission. TEA was consequently allowed to carry out its scheduled services.

In February, just a month after Sabena's Godfroid announced the airline's reorganization, the race to find a partner for Sabena began in earnest. The Belgian government desired to lower its stake in Sabena from 51% to 26%. New investors were permitted to purchase up to 49% of Sabena, with the remainder going to semi-government institutions. Godfroid claimed that discussions with British Airways, American Airlines, and a Southeast Asian airline were underway. The Belgian TEA was another suitor. Joining TEA would seem surprising, considering TEA's complaint to the European Commission about Sabena's unfair tactics. Furthermore, SAS was approached by TEA to make a joint offer for Sabena. That was an interesting move considering Sabena snubbed SAS in a previous merger attempt. An investment company was to be set up by TEA named TEA Invest, primarily to acquire a stake in Sabena. TEA would acquire 45% of the shares and obtain 51% of the voting rights. For its part, SAS would acquire a 25% stake. Above all else, TEA had an advantage in acquiring a stake in the flag carrier because it was a Belgian airline, and the Belgian Government wanted Sabena to remain Belgian.

In September, Sabena signed a partnership agreement with Air Canada for flights between the two countries. Under the 'Codeshare Agreement' terms, Air Canada would have a set number of seats on Sabena's airliners. This deal allowed Air Canada access to Brussels, but the Canadian airline also benefited from the Sabena route network's connections within Europe and Africa. Until October 26, 1991, Sabena's Airbus jetliners would fly the Brussels-Montreal route three times per week. Then, for the winter season from October 27, 1991, the offer would be increased to four weekly flights, including an additional flight on Saturday. Furthermore, flights to Toronto, suspended since March 1991, would be restarted.

With the financial turmoil at Sabena and the collapse of the SWA operations, the carrier's schedules were also affected. The Belgian company maintained four weekly flights from Brussels to Kinshasa, Congo, and established a new route to Burkina Faso. However, there were cuts to Sabena's US operations in the North Atlantic sector, with New York flights minimized to one per day from two. Flights to Boston and Chicago were still available, but not to Atlanta. Sabena discontinued that line. Due to the Tokyo flights now taking the Siberian route rather than the Polar route, Anchorage was removed from the network in East Asia. Toronto, Canada, was also eliminated. Although Singapore was no longer on the itinerary, flights to Bangkok were maintained in Southeast Asia. Middle Eastern destinations were preserved. Although Larnaca and Cairo were dropped from the network, Istanbul and Tel Aviv services were resumed.

1992

Sabena announced in January that for the 1991 financial year, it recorded an operating result of over BFR 2 billion for the first time in its history. The carrier's profit was BFR 45.8 million. Although it was a good result for the airline, especially after the savage losses it had made in the previous two years, 1989 and 1990, it did not mean the airline was financially back on track.

They say hope springs eternal, and Sabena did have reason to have hope. One of the largest airlines in Europe and the world, an airline with a great history, Air France, became interested in getting involved with the Belgian flag carrier. The French giant wanted to buy a 37.5% stake in Sabena. Discussions had been going on for over six months, with Sabena's board sweating for an answer from Air France. Unfortunately, it was not all up to Air France; the Belgian Government's collapse held up the French carrier's plans in November 1991, which caused some political instability. A new Government took office, and a deal with Air France was at the top of the agenda. Not so fast. There was a second hurdle. Political opponents from Belgium's Flemish population believed that French would become the primary language at Sabena. Some Flemish organizations argued that the Air France deal should be scrapped in favor of Flemish ownership, claiming that the French flag carrier would be paying a pittance to move in and dominate Brussels Airport. Unfortunately for the Dutch-speaking Belgians, their calls for canceling the merger with Air France would fall by the wayside, as the national airline's survival was at stake.

Moreover, air France was willing to inject funds into Sabena to keep it flying, which the Belgian Government could not do. In mid-March, the new Belgian government-guaranteed Sabena's continued operation by providing the airline with a significant loan, while legislation allowing the proposed partnership with Air France was pushed through. The Belgian government then took the first move in paving the way for an Air France alliance by framing a bill to allow Sabena to become a public limited company. On March 27, the Chamber of Representatives gave its approval, and Sabena was all ready to become a part of Air France. In early April, the shareholders approved the capital increase. Finally, on April 10, Sabena's president, Pierre Godfroid, and Air France's chairman, Bernard Attali, put their signatures to the cooperation agreement, which finally initiated the long-awaited alliance between the two flag carriers.

Once the dust had settled, the Belgian Government retained a 61.8% stake in the newly formed Sabena. Furthermore, the Government invested BFR 9 billion in Sabena with the European Commission's consent. This investment came after the government had paid BFR 10 billion to keep the struggling airline flying in 1991.Air France brought in an additional BFR 4 billion, allowing it to acquire a 37.5% interest in Sabena. Four Belgian financial institutions held the remaining 0.7%, contributing BFR 2 million. Sabena's share capital was expected to increase from BFR 16.1 billion to BFR 22.1 billion by 1995.

Photo 102. Airbus A310-222 'OO-SCB' at Brussels Airport. Delivered on March 30, 1984. It was sold and delivered to FedEx in November 1997, with registration ' N452FE'. (Photo from 'www.douglasdc8.com')

Photo 103. A310-222 'OO-SCB' on final approach. (Photo by Udo K. Haafke)

Photo 104. Sabena A310-222 'OO-SCA' at London Heathrow in May 1985. Delivered to Sabena on February 23, 1984. Like OO-SCB, it was sold to FedEx in November 1997, with registration 'N451FE'. (Photo by Richard Vandervord)

Photo 105. Sabena's third A310-222, 'OO-SCC' at London Heathrow. Delivered on March 3, 1987. It was withdrawn from use on May 17, 1997. (Photo by Richard Vandervord)

Regarding the Walloon (Dutch-speaking) Belgians, Air France's Chairman Bernard Attali, who was now on Sabena's Board of Directors emphasized Air France's Belgian identity and that everybody at Sabena would respect the Flemish element. He said that Air France's goal was not to swallow up other airlines but rather to work together closely. Attali suggested that commercial alliances were unavoidable in intense competition, especially from the US and Asian carriers. Sabena's journey to find a partner had been successful. Time would tell if it would last.

In October, the European Commission approved the cooperation agreement negotiated by Air France and Sabena in April. The Commission required Sabena and Air France to open up part of their routes between Belgium and France, Europe and Africa, and several non-European Community and European Free Trade Association (EFTA) nations to rival airlines. The sharing of flights was done to keep Sabena and Air France from dominating certain routes. The final obstacle was removed. Sabena and Air France were now partners, with the blessing of the European Commission. The French national airline had invested more than BFR 400 million in Sabena and intended to expand its involvement in the Belgian flag carrier.

1993

Some changes were made in the schedules and liveries in the alliance between Sabena and Air France. Scheduling the regular and well-timed flights (one flight per hour at peak times) on the Paris-Brussels line was the most spectacular achievement for the two airlines. Flights increased from eight to nine per day on this critical route, while seats increased by 20%. In the African sector, flight coordination to the big continent was put in place and ready to start by 1993. Air France would fly to Bamako, Niamey, Ouagadougou, Cotonou and Brazzaville, and Sabena would service the cities of Bujumbura, Kigali and Kinshasa.

Concerning the image of Sabena, there were significant changes. Sabena's airliners would now have the all-white paintwork as Air France's aircraft, with the Belgian airline's name in sizeable pale lettering along each side of the fuselage and repeated more boldly on a small section at the front part of the fuselage, just above the windows. Sabena's tail logo was also modified, with the curved 'S' being painted white, instead of the traditional blue, while the ball encompassing the 'S' was painted in blue. Gone was the blue cheatline that was part of Sabena since before the beginning of the jet age. Repainting the entire Sabena fleet would take about two years. Sabena also became a 'GIE Air France Cargo Group' member on December 28, 1992. EIG was created on January 1, 1991, by the three major French airlines, Air France, UTA, and Air Inter, to develop the interactions of partner airlines in the sector of air cargo.

Air France seemingly had a substantial impact on Sabena in 1992/93 as the Belgian company managed to turn its heavy loss from 1991 into profit in just one year. Sabena's CEO, Pierre Godfroid, stated that this would not have been possible without the cooperation of Air France. Sabena's profit came in at a modest BFR 6 million, compared to the loss of BFR 2.44 billion in 1991. According to Godfroid, Air France's injection of the BFR 400 million for the 37.5% stake in Sabena resulted in a saving of BFR 500 million on management costs for Sabena. The Sabena CEO also stated that Air France and Sabena had started working closely to maintain the aircraft, catering, and computer reservation systems.

At a press meeting, Air France's CEO Bernard Attali was overjoyed by the news of Sabena's profits in 1992. He said that under Mr. Pierre Godfroid, Sabena was able to take the necessary measures and that those who doubted Sabena's recovery capabilities and the relevance of the investment made by Air France could be reassured. Attali also welcomed Sabena's new visual identity, signifying that

Sabena belonged to the Air France family. However, someone should have told Mr. Attali that one swallow does not make a summer. The airline was the problematic Sabena, after all.

Photo 106. This Sobelair 737-229, 'OO-SBT,' is painted in all white colors in the post-Air France/Sabena alliance. (Photo from 'www.douglasdc8.com')

In June, Sabena celebrated its 70th anniversary. For the occasion, the Belgian airline planned a large Sabena family party for the 22nd and 23rd of May, which was just for its employees. The 9,904 people invited to the open house weekend took part in an event titled 'The four continents on which Sabena is present.' There were also several exhibitions which traced the company's history and its activities. One of the highlights of the occasion was the Belgian pianist Francois Glorieux's performance of a composition written in honor of Sabena and composed primarily for the airline's 70th anniversary.

Unfortunately, the celebrations were cut short in July, as Sabena was in a new crisis. Heavy first-half losses by the company prompted the Belgian national carrier to launch a cost savings program, including pay cuts of up to 15%. The unions were up in arms and did not want to hear a word of salary reductions. On August 4, Sabena's Board of Directors met and warned that, without any salary cuts, further layoffs would be inevitable if the labor unions continued to resist. The management also reiterated the necessity of shaving BFR 2 billion off the salaries over the next three years to avoid bankruptcy. The warnings came in the aftermath of a labor vote that rejected suggested salary cuts by 56.2%. Godfroid fulfilled his promise to present his resignation following the ballot defeat, but the board asked him to postpone any decision until August 19. Sabena admitted that earnings in the first half of 1993 fell BFR 1.5 billion short of budget, fueling fear that the airline would complete the year with significant losses totaling up to BFR 6 billion. In response, Godfroid suggested a plan to decrease expenses by BFR5 billion, with enhanced labor flexibility accounting for BFR2 billion, the cutting of routes accounting for another BFR1 billion, and salary reductions accounting for the remaining BFR2 billion. However, in response to the heated union objections, Godfroid agreed to limit salary cuts to BFR 1.1 billion, which the Board of Directors eventually reversed. The airline's two most significant shareholders, the Belgian government and Air France, unequivocally declared that they would not bail out Sabena with new funds. The burden was solely on Sabena's shoulders. Finally, the unions agreed to salary cuts at the last minute. Salary cutbacks of 2% to 17% were proposed in the new proposal, although they would be distributed based on income and over two years. The ABPNL pilots' union, which did not sign the agreement, was still negotiating with Sabena's management.

Sabena hit the road again looking for new partners and found an interested party in the form of American Airlines in October. The American carrier had commissioned experts to investigate if Brussels could become a unique European central point. From which American Airlines could begin flying with Sabena under shared flight numbers and ground management. The Belgian company was enthusiastic and wanted to start working together right away, but American was more cautious. Moreover, Sabena was already sharing the Brussels-Atlanta route with Delta Air Lines. American was looking for a European partner because its closest competitors in the US had already established links with European carriers. The American behemoth was also in talks with Lufthansa, which instead chose American Airline's US rival, United Airlines.

Sabena planned to expand its agreement with Delta Airlines the next month. In mid-January 1994, the Belgian airline would run the Brussel-New York route in collaboration with Delta. Since June 1993, the two airlines have flown to Atlanta together, with Sabena providing the aircraft and crew. Delta was also a partner of Swissair, which went on to form the 'Alcazar' alliance with KLM, SAS, and Austrian. Following the demise of the 'Alcazar' group, Sabena and Swissair would develop closer ties. The 'Alcazar group was formed in 1993 when the national airlines of four European countries had a vision of uniting and forming a new airline that would be a giant in the civil aviation world. The four airlines combined had 288 aircraft, transported approximately 325 million people in 1992, and conveyed 858,000 tons of cargo and mail. They employed 70,000 people and flew to 400 destinations throughout the world. Swissair, SAS, KLM, and Austrian Airlines were the four airlines. In March 1993, two of the airlines, Swissair and Austrian Airlines, were concerned that as part of the contract to join 'Alcazar,' they would have to give up their distinct identities. With KLM playing a vital role in the prospective alliance, the Dutch airline pushed for a swift conclusion to the talks. Its frustrated Director, Leo van Wijk, stressed that the four airlines must stop wasting time and decide as soon as possible, or the discussions would become bogged in the details. The 'Alcazar Project' collapsed due to a difference in which US carrier to invite to the group. KLM preferred Northwest Airlines, as they were partner airlines, while Swissair chose Delta Airlines, with which it had a good relationship. On November 6, 1993, the Alcazar group met in Stockholm to try and get the airlines to agree on a US partner carrier. Unfortunately, the two antagonists, Swissair and KLM, stuck to their airline of choice, causing the talks to fail. Two weeks after the November 6 meeting, the Alcazar Project was over. The end of the 'Alcazar' project opened the door for Sabena to team up with Swissair.

The Sabena-Swissair Alliance Part 1

Sabena's results, released in April, were nothing to write home about. The Belgian carrier posted another loss of BFR 4.5 billion for 1993. The loss, which compared to the modest BFR 6 million profit Sabena made in 1992, came after a challenging year. The causes for the failure were mainly external. There was the ongoing political upheaval in the Congo and other central African regions, which meant Sabena could not serve those lucrative routes for six months, resulting in a BFR 1.5 billion loss. Furthermore, the airline's revenues were reduced by BFR 1 billion due to the depreciation of crucial European currencies against the Belgian franc. Sabena's CEO, Pierre Godfroid, said productivity had grown by 45% in 1993, even though the staff had been downsized from 12,000 to 9,500. In addition, passenger numbers increased by 16%, while volume on the European route network increased by 22%. Godfroid believed that cooperation with Delta Air Lines was possible if the American airline could be persuaded to shift its European hub from Frankfurt to Brussels due to the 'United Airlines-Lufthansa' cooperation deal. Godfroid said that Delta might be interested in buying a stake in Sabena if Air France decided to sell its 37.5% stake in the company. However, that was not on the cards, as Air France was committed to keeping its stake in Sabena.

The search for a partner continued for the Belgian airline. Then in June, there was interest in Sabena from another European carrier that was similar in size to the Belgian company. It was the Swiss national airline, Swissair. The Swiss airline was also looking for partners, and it became interested in Sabena after the collapse of the Alcazar Project in 1993. Thus, Swissair and Sabena began talks in June 1994 to form a collaboration agreement. The opportunity for Swissair to invest in Sabena arose after Air France experienced severe financial difficulties and decided to sell a 24% ownership in the Belgian airline. Swissair was looking for a partnership with Sabena for its own reasons. The Belgian carrier was from an EU country, and Swissair, which was not from the EU, would be able to realize its long-awaited dream of having unrestricted route access to any destination in the European Union by forming a relationship with Sabena. As for Sabena, the possibility of reaching a deal with Swissair was also substantial for Sabena, as the Swiss carrier announced that it would provide funds to help save Sabena. Swissair was aware that Sabena was being courted by itself and by some other airlines from Europe and America. Swissair hoped it could woo Sabena and climb to the top of the Belgian carrier's preferred partner list. There was one problem, Air France would have to sell its whole 37.5% stake in Sabena before Swissair could make its move on a partnership with the Belgian carrier. Most airline executives in the civil aviation industry believed that Swissair had little chance of convincing Air France to sell its whole stake in Sabena. It looked as if the airline executives were right when in July, despite Swissair's interest in acquiring Sabena, Air France reaffirmed that it had no intention of selling its 37.5% stake. The French flag carrier also stated that it would honor the initial terms of its Sabena commitment and pay the final BFR 1 billion portions of its investment, which happened in August. The payment of the final installment to Sabena put to rest fears that Air France may be forced to end the Sabena agreement due to its own restructuring plan. However, the European Commission approved Air France's plan, allowing the French carrier to make the final payment. Swissair, for its part, was quietly confident that Air France, because of its dire financial circumstances, would eventually relinquish its interest in Sabena.

Sabena wanted to purchase up to 12 British Aerospace BAe 146 regional jets to put on its routes within Europe. Delta Air Transport (DAT), Sabena's 79% owned Antwerp-based commuter subsidiary, had been operating 4 of the type since 1992. Sabena's CEO Pierre Godfroid said he was pleased with the quietness of the BAe 146, both inside and out. The noise-sensitive Berlin Tempelhof Airport, where the BAE 146 was the only jet-powered aircraft allowed to land, was one of the routes the Bae 146s would operate once they joined Sabena.

1995

The Sabena-Swissair Alliance Part 2

The union between Sabena and Swissair was not consummated in 1994, as some loose ends needed to be fixed. In January 1995, Swissair's President, Otto Loepfe, said that Swissair only wanted to take a position in Sabena if it could acquire a majority stake in the Belgian airline within five years if the European Union's (EU) rules made that feasible. Initially, Swissair said it only wanted a 49% stake. The new claim took the Sabena team by surprise, but the Belgian government did not reject the request. Sabena needed a BFR 12 billion injection to survive. Belgian transport minister, Elio di Ruppo, suggested that Swissair was willing to help increase Sabena's capital. Swissair was the only party that stated it would pour money into Sabena. However, the largest shareholder in Sabena, the Belgian Government, could not put that amount of money on the table. Moreover, Sabena's partner at the time, which also had a 37.5% stake in Sabena, Air France, was also not considered capable of providing the funds, as its financial position was precarious. Swissair knew that and upped the ante by asking for a majority stake in the Belgian flag carrier.

In mid-January, Sabena said it wanted to register, or 'flag out,' its entire fleet and all 480 pilots to Luxair of Luxembourg to take advantage of the favorable rates of taxation given to the Luxembourg airline by the Luxembourg Government. The social charges in Luxembourg were almost four times lower than in Belgium. Luxair's management told Sabena that it would promptly decide whether it could approve Sabena's request. As a sweetener, the Belgian airline offered to participate in injecting capital into Luxair. In the proposed plan, Sabena's 480 pilots would be employed by Luxair or a Luxembourg company yet to be established. Luxair, or the new company, would then 'hire' the pilots to Sabena. As for Sabena's aircraft, they would be given Luxembourg registration. It was a far-fetched gamble by Sabena, but it had to resort to such measures to stave off bankruptcy. If Sabena and Luxair had reached an agreement, all would not be over. The Governments of Belgium and Luxembourg also had to consent to the 'flagging out' operation. The 'flagging out' of the Sabena fleet to Luxembourg did not affect the talks between the Belgian airline and Swissair. The answer was given by Luxair to Sabena only a few days later.

The reply was not the one that Sabena wanted to hear. Luxair's Board of Directors said they disagreed with Sabena's request, as too many risks were associated with the Belgian carrier's plans. In addition, the Luxembourg airline wanted to strengthen further its ties with Lufthansa, which had a 13% stake in Luxair. Ultimately, Sabena's gamble did not pay off.

Photo 107. Sabena BAe-146, 'OO-DJJ', at Brussels Airport. Delivered to DAT on June 28, 1991. The aircraft is in the Sabena scheme as it was operated on behalf of Sabena. (Photo from 'www.douglasdc8.com')

Photo 108. This BAe RJ85 was initially delivered to DAT on June 27, 1996. It is seen here in Sabena colors, with the 'Qualiflyer Group' logo on its belly. After the collapse of Sabena, the aircraft was transferred to Brussels Airlines, where it flew until December 4, 2011. (Photo by Richard Vandervord)

Photo 109. This is an Avro RJ100 of Sabena, 'OO-DWH,' delivered to DAT on December 16, 1998. It was transferred to Brussels Airlines on February 15, 2002. (Photo by Eduard Marmet)

Photo 110. This BAe 146-300 was initially delivered to DAT on August 14, 1990. It is seen here at Manchester Airport on May 28, 2003, in Brussels Airlines colors after it was appropriated by that airline after the collapse of Sabena. The DAT and Sabena logos can still be seen to the right of the front entrance door. (Photo by Richard Vandervord)

In February, Sabena took another hit when Swissair halved its offer to Sabena. The proposal was reduced from BFR12 billion to BFR 6 billion. That meant that Sabena would have to find the other BFR 6 billion from other sources. In March, the Belgian Government again came to Sabena's rescue. The Belgian state's exemption of Sabena from social legislation for approximately BFR 800 million of the airline's annual costs helped pave the way for its proposed cooperation agreement with Swissair. The positive for Sabena was that, in the new terms, Swissair would take a 49.5% interest in the Belgian airline. Thus, there would be no question of a takeover of Sabena by Swissair. The two airlines would remain independent. However, due to its substantial participation in Sabena's capital, Swissair would have a big say in Sabena's policies. In April, to push for a conclusion in the negotiations, the Belgian government approved the far-reaching collaboration between Sabena and Swissair.

In May, the Belgian Government initially thought that a reduction in social security charges, which would save Sabena BFR 800 million per year, would be enough. That plan, however, encountered resistance from the European Commission, which considered it to be contrary to EU competition rules. Consequently, a new project was done by the Belgian Government. The state would limit Sabena's social charges to BFR 300 million. In addition, as a shareholder of Sabena, the Belgian state would increase its contributions to the airline. This increase saved Sabena almost BFR 130 million in interest.

The union of Sabena and Swissair was completed on Thursday, May 4, 1995, when top representatives from both airlines and Belgian Transport Minister Elio Di Rupo signed the agreement in Brussels. According to the terms of the deal, Swissair paid BFR 6 billion for the holding and invested an additional BFR 500 million in non-voting certificates. Furthermore, the Belgian government pledged to invest BFR 1.5 billion in the airline over the next ten years through the 'Societe Federale d'Investissement' (SFI), a federal Belgian finance company. Swissair, for its part, had to offer a convertible loan to the Belgian government of BFR 4 billion, allowing SFI to buy back Air France's Sabena stake through the Finacta consortium. In 1992, that consortium paid BFR 6 billion for a 37.5% stake in Sabena as part of the Belgian airline's survival strategy. The remainder of the Finacta stake was held by Belgian investors, who retained an interest in Sabena.

Then, on July 19, the European Commission fully approved the cooperation between Swissair and Sabena. Thus, Swissair officially took a 49.5% stake in Sabena. The EU added that air traffic between the two countries had to be liberalized; otherwise, the Belgian-Swiss connections would create a clear monopoly position. However, the EU had some conditions. A monopoly could arise on flights between Switzerland and Belgium. Consequently, the number of flights of the two airlines between their two countries could only be increased up to 25%. Moreover, Switzerland had to allow four EU airlines the opportunity to fly from an EU country via Brussels to Switzerland. Furthermore, Sabena had to remain nominally in Belgian ownership. However, Swissair was aiming for public participation in the long term, which would only be possible if the EU conditions became relaxed. Due to Switzerland being a non-EU country, Swissair was not considered an EU airline, even with its participation in Sabena. Swissair's flight rights to EU states still had to be negotiated on a case-by-case basis.

Sabena's 1994 performance results were excellent. Sabena transported 4.25 million passengers, 17% more than in 1993, marking the carrier's highest increase in passenger traffic ever. From that figure, 3.56 million passengers were transported across Europe's route network, nearly half a million more than in 1993. Sabena explained that the significant increase was due to the success of the Brussels hub. In addition, Sabena's occupancy rate increased in

1994, mainly on flights to Africa and Japan. In the financial area, Sabena reported a positive operating result of BFR 810 million in 1994. However, the net result remained negative due to the high financial charges. As a result, Sabena incurred a loss of BFR 4.5 billion in 1993, including an extraordinary reorganization charge of BFR 1 billion. However, from February 1994 onward, Sabena's turnover exceeded expectations and reached BFR 54 billion in 1994, increasing by 7% from 1993. The results improved due to the following factors: significant traffic growth, particularly on the European network, the full effect of the recovery measures implemented in August 1993, and good results of Sabena's subsidiaries.

In late November 1995, Sabena was hit by one of the worst strikes on record. In response to Sabena's management's unilateral termination of all collective labor agreements. The unions were outraged when Sabena's CEO, Pierre Godfroid, announced the move on November 27. Consequently, the unions organized a 24-hour strike by the entire Sabena staff. Pierre Godfroid unilaterally canceled the collective agreements of Sabena's 9,500 employees. The contracts included refunds, travel charges, and the all-too-common Belgian thirst for food coupons. The complications occurred during Sabena's discussions for a social pact. Mr. Godfroid promised to create 1,000 jobs over the next three years, but it would require a significant effort from the staff, such as job flexibility, a 5% increase in working time, and a 3-year wage freeze. The unions were willing to discuss the flexible working hours and the increase in work time, but they declared that the assault on wages was unacceptable. Sabena's management immediately left the negotiating table after the unions became more vehement in their administration's criticism. Sabena had to cancel all 140 flights from Brussels Airport during the one-day strike.

1996

In mid-February, Sabena was paralyzed by another strike. This time, the Belgian pilots' union, ABPNL, opposed Sabena's hiring of four pilots of its regional subsidiary, DAT, to fly the Belgian flag carrier's new Avro RJ-85s. The latest confrontation was the third in a series in November 1995, when unions walked out on talks on a business plan that included employment flexibility, a three-year salary freeze, and a 5% rise in working time. In addition, Sabena's management stated that a percentage of future dividends would be set aside for airline employees. Sabena's CEO, Pierre Godfroid, subsequently revoked all labor agreements, the first time this had happened in Belgium. The unions went on an immediate 24-hour warning strike, ridiculing Godfroid's promise to create 1,000 new 'low-cost' jobs. Sabena's management had frequently yielded to the unions following such warning strikes in previous years. However, Sabena's new management team reacted aggressively, claiming that Sabena was losing BFR 100 million per day of industrial action and that the airline would recover the losses through workforce reductions.

Regarding the third pilot strike, DAT's pilots were to be paid 35% less than Sabena's pilots, which, according to ABPNL, violated a previous agreement that stated DAT pilots flying Sabena routes would be paid the same as Sabena's flight crews. Sabena's management responded that the RJ85 was not a BAe-146 and that the previous labor agreement did not apply. Moreover, the airline argued that paying higher salaries to DAT aircrews would make the services uneconomic. In addition, the company insisted that a low-cost operation was critical to the airline's increasingly successful hub-and-spoke operation, much of which was operated by DAT. The strike was resolved at the end of May when an agreement was reached specifying that low-cost DAT would initially employ pilots, but with the possibility of advancement into the main Sabena fleet.

Swissair's involvement with Sabena intensified in February. Paul Reutlinger, a member of the Swissair Group, was appointed CEO of Sabena. Mr. Reutlinger was previously the President of Switzerland Tourism, which he had held until his move to Sabena in April 1994. Paul Reutlinger took over Sabena's operational management and replaced CEO Pierre Godfroid, who had resigned just a few days earlier. In essence, control of the Belgian carrier had now been transferred to Swiss hands. Mr. Reutlinger's assignment was to get the Belgian airline out of its financial trouble. He had to restructure Sabena's management to comply with stricter financial rules, which a subsidized national airline like Sabena had little experience. When he started his new job, Reutlinger's responsibilities included submitting a restructuring plan for Sabena's Management Board to Sabena's Board of Directors and arranging negotiations with Belgian unions.

Photo 111. The new Swiss CEO of Sabena, Paul Reutlinger, is seen here second from left, with three other airline CEOs: Swissair, Philippe Bruggisser on the left, Austrian Airlines, Mario Rehulka second from right, and Delta Air Lines (USA), Leo Mullin, on the right, on February 1, 1997. (Photo by ETH Zurich)

Sabena's previous president, Pierre Godfroid, was forced out during the bitter union disputes that engulfed Sabena in early 1996. Consequently, in June, Sabena's new partner Swissair laid out details of new cost-cutting targets and fleet rationalization plans that would have to be implemented by the Belgian airline. Sabena's new Swiss CEO, Paul Reutlinger, was not one to let the unions take over Sabena. To bring Sabena back to profitability by 1998, the Belgian airline needed to cut BFR 2 billion in labor expenditures through greater production and new working conditions. Improvements in operations were expected to save another BFR 2.7 billion. A deal with the unions was expected to be reached by October. However, Reutlinger warned that if the unions did not agree to the savings, Sabena would be forced to reduce salaries by 15% or lay off 1,700 employees. Three of Sabena's unions agreed to meet for discussions, while the fourth declined. One of Sabena's priorities was to halve unit costs and boost load factors to 57% in Europe and 67% on intercontinental routes.

Furthermore, in fleet rationalization measures, the carrier's two McDonnell Douglas DC-10s and three Airbus A310s would be phased out, leaving Sabena with three Boeing 747s and four Airbus A340s. Another option looked at by Reutlinger was the transfer of all 23 RJ85 regional jets and their crews to Swissair subsidiary Crossair and then having DAT wet-lease them back. Doing so would save BFR 200 million due to the lower tax rates in Switzerland. The Belgian airline's maintenance business, Sabena Technics, would concentrate on specialty work, such as overhauling Boeing 727s and 737s, but its wide-body operations were likely to be cut. Sabena's catering operation would be combined with Swissair's rapidly expanding 'Gate Gourmet' business, while the company's faltering air cargo operations were primed to be taken over by its Swiss partner unless performance improved. The overall financial target set by Swissair was to improve Sabena's return on capital to 4% by 1998, rising to 8% by 2000. Swissair was now firmly in charge at Sabena.

Photo 112. Sabena Airbus A330-301 on approach to Zurich-Kloten Airport on February 27, 1999. It was an ex-Air Inter aircraft. Air Inter merged into Air France in 1997. Consequently, this A330 was purchased by Sabena and delivered to the Belgian airline on November 12, 1997, with Air Inter's registration 'F-GMDC.' It was reregistered as 'OO-SFO.' (Photo by Aero Icarus via Wikimedia)

Photo 113. Sabena Airbus A330-223, 'OO-SFT,' taxiing at Brussels Airport. Sabena leased this aircraft from its partner Swissair. It was delivered to the Belgian airline on February 3, 2000. It is seen here in Sabena's final colors. (Photo by ETH Zurich)

Photo 114. This is Sabena A340-211, 'OO-SCW', at Zurich-Kloten Airport on August 23, 1998. Purchased from Air France with French registration, 'F-GNIB.' Delivered to Sabena on June 25, 1996. Six days after this photo was taken, the airliner's right main landing gear burst and caused the aircraft to veer off the runway. There were no injuries. (Photo by Aero Icarus via Wikimedia)

Photo 115. Sabena A340-311, 'OO-SCZ.' Delivered to Sabena on April 15, 1997. After the collapse of Sabena, it was stored at Chateauroux, on February 20, 2002, before being acquired by Lufthansa on February 28, 2003. (Photo by Contri via Wikimedia)

The Swiss CEO Paul Reutlinger unveiled a new plan on August 21st, dubbed the 'Horizon 1998' project. Sabena would be able to return to profitability within two years under the new scheme. After a meeting between management and the unions on September 10, all parties believed they had progressed on an essential issue. They both looked to be on the same page regarding early retirement. This arrangement was expected to result in the departure of 500 of Sabena's 9,600 employees. This agreement, however, required the consent of the Belgian Ministry of Employment. The Belgian State, as Sabena's second shareholder after Swissair, was likewise concerned about the smooth running of the airline. Management and labor unions were also negotiating methods to boost the company's productivity. The 'Horizon 1998' plan provided for greater flexibility from staff in the organization of working conditions. Reutlinger planned to cut personnel costs by BFR 1 billion. However, Sabena was still looking for savings of around BFR 2.7 billion in other areas, including inside management. Finally, in late September, the unions came through. The two sides reached an agreement on five ways to cut Sabena's excessive salary bills. These included: 1) a pre-retirement system that allowed employees to leave at the age of 52, 2) voluntary resignations, 3) the implementation of temporary jobs, 4) the ability for employees to take unpaid leave, and 5) job flexibility. The measures would result in a 1,270-person decrease in the workforce, which would fall short of the BFR 2 billion labor-cost reduction target. New wage reductions were viewed as unavoidable to make the necessary savings. Meanwhile, Reutlinger stated that salary reductions were not considered a priority for Sabena.

In fleet matters, Sabena's fleet expansion was in full swing, with the European Investment Bank (EIB) partially financing the enlargement of Sabena's fleet. EIB granted Sabena a loan of BFR 3.6 billion. The first part of BFR 1.2 billion was made available to the Belgian carrier on November 8. Sabena's President Paul Reutlinger and EIB Vice President Rudolph de Korte signed the agreement in Brussels. The funding would contribute to the renewal and extension of Sabena's short and medium-haul fleets.

Sabena's management placed an order with the UK aircraft manufacturer, British Aerospace, for 23 Avro RJ85s in September 1995. By November 1996, 13 Avro RJ85s had already been delivered and put into service, with one more to be delivered before 1996. Sabena then converted its nine outstanding orders for Avro RJ85s to orders for the larger BAE Systems Avro RJ-100 model, which offered accommodation for 95 passengers. The RJ-100s would supplement the 12 earlier 82-seater RJ85s already operated by Sabena's regional subsidiary, DAT. Deliveries of the RJ100s would be spread through to 1999. Sabena's goal was to form a modern and environmentally friendly fleet. Paul Reutlinger stated that Sabena would limit itself to two types of airliners for its European network in the future, the BAe Avro RJ100s and the newer Boeing 737s. However, the news was bad for the older, noisy 737-200s. Sabena's expansion plan called for them to be replaced sometime after 1998. The rationalization of Sabena's long-haul fleet was also on the agenda. In its long-haul services, Sabena aimed to reduce the three long-haul models, the 747s, Airbus A340s, and DC-10s. The return of two leased DC-10-30s back to Lufthansa was scheduled for 1998. That would leave Sabena with only three 747s and five A340s, two of which were leased from Air France. Sabena was to have just a single airliner type in its long-haul fleet by the year 2000.

1997

In 1996, Sabena was very deep in the red. That year, its consolidated loss was five times higher than the loss in 1995, making it the worst result ever in its history. In 1996, the carrier recorded a BFR 8.87 billion deficit, compared to the BFR 1.66 billion shortfalls in 1995. Moreover, the airline's net loss was BFR 8.32 billion, compared to BFR 1.99 billion the previous year. It also recorded a BFR 4 billion operational loss, up from BFR 3 billion in 1995. Furthermore, the company's expenditure was BFR 64.67 billion, compared to BFR 59.35 billion in 1995, with a negative cash flow of minus 2.4 billion. Cash flow was favorable in 1995, totaling + 1.5 billion. The only positive to come out of 1996 was Sabena's operating revenue, which increased to BFR 60.63 billion in 1996, up from BFR 59.35 billion the previous year. Overall, financially, it was a disastrous year on a grand scale. The heavy losses in 1996 were attributed to the wildcat strikes early in 1996, which had a lingering impact on passenger bookings. There was also the money being spent to underwrite Reutlinger's restructuring measures.

Reutlinger remained positive and believed that Sabena's results would improve substantially in 1997, which would allow the Belgian airline to stay on course to reach the target of breaking even by 1998. However, he warned that Sabena had to prove, as soon as possible, that it was capable of breaking out of the red. Sabena's CEO said that due to the poor performances of the Belgian carrier, the board of directors of Swissair had authorized the Swiss airline's management to write down its BFR 6 billion investment in Sabena in the second quarter of 1997. In terms of the agreement, Swissair could cut its losses and withdraw from the Sabena union if Sabena did not record positive financial results by 1998. Even if Zurich did not set an ultimatum on a specific date, it was clear that the future of the Sabena-Swissair union was going to be rocky.

In a conference in Brussels in January, Reutlinger claimed that Sabena would meet Swissair's target of returning to profitability by 1998. To facilitate this, he outlined five significant goals: 1) Cutting down the costs of ton-km from BFR 32 to BFR 27. 2) An increase in load factor to 57% in the European sector, from the current 1997 level of 52%, and a 67% increase elsewhere on the airline's route network. 3) An increase in the share of business traffic from 32% to 37%. 4) A decrease in connection time to an average of 27 minutes. 5) An improvement in punctuality by 1998 to boost Sabena from fifth to third place among major European airlines.

In March, Sabena introduced its summer timetable. However, the 1997 timetable was different from all others before it. In the new 1997 timetable, Sabena introduced a series of tailor-made improvements for business passengers. The image change was part of the 'Horizon 98' plan, conceived by Sabena's Swiss CEO Paul Reutlinger, to propel the airline back to profitability. The new Sabena brand was created to meet customer expectations better. With the new image, Sabena suddenly became a European business airline. The cabin crew was more attentive to the passengers, from purchasing their tickets until arriving at their destination. The Belgian airline introduced an ongoing education program in January to improve the helpfulness of its staff. In addition, Sabena wanted to distinguish between economy and business class by offering a more personalized service with symbolic aspects like a welcome drink for the business passengers and more appealing meals. The new services began on March 30. In addition, Sabena drastically improved its punctuality in 1997, jumping from 11th to 1st place in the rating of the most famous European corporations in December 1996 and January 1997. Furthermore, Sabena's management, which was dissatisfied with the baggage check-in, transit, boarding, and disembarkation procedures at Brussels airport, engaged an airport management business to handle those tasks.

For the 1997 summer schedule, Sabena flew to the following destinations: London Gatwick, Nantes, Moscow, Palma, Beirut, and Cincinnati. Gatwick Airport was serviced together with Richard Branson's low-cost airline, Virgin Express, with which Sabena also shared flights to London Heathrow, Barcelona, and Rome.

In a surprising move in November, Sabena acquired a stake in City Bird, a new Belgian start-up airline. The two airlines announced a collaboration agreement to cover additional long-haul routes. Sabena paid around BFR 18 million in shares and options for a 12% ownership in the new airline. When the flights began on May 1, 1998, the two carriers said they would be code-sharing on flights to Montreal, New York, and Sao Paulo. According to an agreement reached with the unions, the services would use two City Bird aircraft painted in Sabena colors and flown by Sabena pilots. City Bird estimated that the code-sharing flights would generate BFR 2.4 billion in revenue for the airline in 1998.

The year 1997 was a busy year for Sabena's fleet. In January, Sabena placed a firm order for one A330-200, which would be delivered in September 1998. In addition, it planned to lease three A330-200s from International Lease Finance Corporation (ILFC), an airline leasing company. While waiting for the 4 A330s, Sabena rented three ex-Air Inter A330-300s from Airbus Industrie in the interim period. Then, the Belgian carrier retired all of its DC-10-30s on April 27. Sabena initially operated five DC-10-30CFs in the early 1970s, but they were sold off in the early 1990s. Sabena then leased two DC-10-30s from Lufthansa, as its six Airbus A340s were leased to Air France. With the return of the A340s to Sabena in 1997, the 2 Lufthansa DC-30s were returned to the German carrier. Later, Sabena's board of directors decided to acquire 34 A320 medium-haul airliners in a stunning move. It was Sabena's most prominent order in its history. Synergies with Swissair weighed heavily in the decision, as the Swiss airline was already operating 34 A320s. CEO Paul Reutlinger declared that Sabena chose the A320s because Airbus had offered highly favorable terms for the transaction and made significant concessions. Sabena's A320s would go on to replace the carrier's 14 old 737-200s.

Photo 116. City Bird 767-33A-ER, 'OO-CTA' in South Africa. Sabena took a 12% stake in this airline in October 1997. (Photo by Bob Adams via Wikimedia)

The A320s would start being delivered to Sabena from late 1998, while the phasing out of the 737-200s would be completed by 2000.

In essence, the choice of Airbus to provide the aircraft needed for Sabena's fleet ended a dispute between Sabena and the unions. Both Airbus and Boeing were trying to entice Sabena to buy their products. The two rival aircraft makers even carried out demonstration flights for Sabena in Brussels with their latest models, in Boeing's case, the 737-700, and for Airbus, the Airbus A319. The race to get the contract was tight. Sabena's workforce wanted Boeing to get the contract from the outset as Sabena's fleet was primarily made up of Boeing airliners. Moreover, in Sabena's technical department, 'Sabena Technics,' more than a thousand people worked on the Boeing 737 maintenance lines for Sabena and several other airline companies. Hence, the technicians wanted Boeing to get the contract. The unions supported the technicians and did not hesitate to threaten strike action if Airbus prevailed. Due to pressure from the technicians and unions, Sabena's management decided to cancel a press trip to Toulouse, the headquarters of Airbus, at the end of September, in case the unions opted to launch work stoppages. This incident prompted Airbus to invite Belgian journalists to visit the company in Toulouse directly. Boeing added fuel to the fire by declaring 'Sabena Technics' would become a central maintenance center in Europe for Sabena's aircraft and other airlines' aircraft. Airbus ultimately got the deal because it went far further than Boeing. Airbus told the technicians and unions that it would guarantee Sabena's technical center jobs for the next ten years. It would also train the technical staff at its expense and finance the installation of a maintenance line for its aircraft. Boeing had been defeated. The American firm returned home empty-handed. Airbus had won the day.

1998

Sabena signed a partnership agreement with the second-largest airline in France, Air Outre Mer (AOM), covering routes between Brussels, Marseille, Toulon, Nice, and French overseas territories and other far-flung destinations. Sabena and AOM were part of the Swissair group.

Swissair announced in March that it would join forces with its partner airlines to form the 'Qualiflyer Group,' a pan-European airline alliance. Swissair established the Qualiflyer Group to improve collaboration between Swissair and its partners. The group members were: Swissair, Sabena, Crossair, Austrian Airlines, TAP Air Portugal, Turkish Airlines, Air Littoral, and AOM of France. Qualiflyer was essentially a frequent flyer program in which the passenger would accumulate airline points and eventually obtain free flights, among other perks. Balair and LTU were not members of the group, although they did participate in the group's frequent flyer program. According to the group, smaller airlines could only survive in the new aviation world order of deregulation and other complexities if they joined an alliance. With the extended new coalition, the group's route network expanded dramatically from 197 destinations to 294.

Although the idea was good and the motivation was there, the Qualiflyer coalition was doomed. It was made up of smaller, second-tier European airlines, the majority of which were heavily indebted and had to deal with strong unions in their own countries. Furthermore, although Sabena was a top-tier airline, it was also a failing carrier that could not contribute financially to the group. In essence, the alliance only had one stable airline, Swissair. The Swiss airline was the only one in the group with a brand name, high-quality services, a modern fleet, and, most significantly, it was a financially stable airline with outstanding leadership and an excellent management team. The onus was on Swissair to financially prop up the ailing airlines in the group, such as Sabena. Keeping the 'Qualiflyer' group functioning was one primary reason for Swissair's ultimate demise.

Sabena significantly improved its results in 1997, as it recorded an operating profit. Nonetheless, high restructuring costs resulted in a consolidated loss of BFR 2.5 billion for the year. That was not so bad compared to the loss in 1996, which was three times larger at BFR 8.87 billion. Sabena's operating result before financial charges and taxes in 1997 was BFR 616 million. In contrast, the 1996 operating result was minus BFR 3.4 billion. Other positive results included a 16% increase in turnover to BFR 72 billion, reflecting the robust expansion in passenger traffic. Sabena transported 6.7 million passengers in 1997, a 33% increase over the previous year. With this increase, the Belgian company was way above the European airline average of 9%. The rise in Europe was 29.6%, but the airline's long-haul network took top honors by recording a mammoth increase of 48.2% in passenger numbers. As a result, the average occupancy rate of Sabena's aircraft increased from 59% to 66%. This positive result came about because of a helpful economic situation and increased flights on the European flight network.

City Bird launched transatlantic services on behalf of Sabena in April, utilizing two new MD-11s painted in Sabena's colors. The MD-11s were reconfigured to include business and economy class cabins for Sabena's passengers and a separate City Bird cabin for the no-frills flyers. Sabena was in charge of publicizing the new flights from Brussels to New York, Montreal, and Sao Paulo. The Sabena and City Bird collaboration expanded on May 6 when the two airlines launched two flights per week from Brussels to Montreal on City Bird's MD-11s. The new service marked the return to Canada for Sabena, which had suspended flights to that nation in 1993.

Sabena's partner airline, Delta Air Lines of the US, became more involved with Sabena in 1998. The American carrier commenced code-sharing flights with Sabena on the daily Brussels-

Oslo route operated by the Belgian carrier's 737-329s on March 13. Then, on April 15, Delta Air Lines began code-sharing flights with Sabena on the two City Bird MD-11s on the daily Brussels-Newark route. Sabena's code-sharing agreement with Delta was expanded further on May 15 to include 4 of the Belgian carrier's five lines to the United States. The Belgian airline opened a second daily Airbus A340-311 Brussels-Atlanta service and increased the flights from Brussels-Cincinnati from 5 per week to daily flights. Meanwhile, on the Brussels-Boston route, it started operating five weekly flights. Finally, on the Brussels-Chicago line, it boosted capacity by replacing its Airbus A340-311s with the higher volume Boeing 747-329C, which operated a daily service to the windy city, except on Saturdays. One month later, Delta Air Lines participated in two other code-sharing flights with Sabena, both in Africa, on June 1, on the Brussels-Conakry (Guinea) line and five days later on the Brussels-Banjul (Gambia) route. Both services were flown on Sabena's airliners. The code-sharing teamwork with Delta continued into September when on the sixth of that month, the American airline joined Sabena on the Belgian carrier's weekly A340-311 service Brussels-Lomé (Togo).

1999

In early 1999, Sabena achieved some positive news. For 1998, the airline recorded a net profit of BFR 703 million, its first since 1958. It also posted a turnover of BFR 86.84 million, which was a 21% increase over the previous year. Its recovery resulted from extensive internal reorganization and establishing a hub-and-spoke operation based in Brussels. The airline also saw a substantial increase in passenger numbers, with 8.75 million passengers carried in 1998, a 27% increase over 1997's 6.87 million. It was a tremendous result, considering Sabena was no longer a great intercontinental airline. The heavy losses over the years trimmed a lot of destinations off the carrier's route network, which essentially turned Sabena into an intra-European carrier, with 83% of all passengers traveling within Europe. Sabena's overall load factor in 1998 was 67.5%, with the European figure increasing by 4.5% to 63% and the long-haul figure decreasing from 72.2% to 70.9%. Long-haul flights surged by 47% in 1999 and accounted for 17% of all passengers.

Sabena and El Al inked a code-sharing agreement covering flights between Brussels and Tel Aviv in January, and in April 2000, both airlines increased their frequencies. In addition, the deal allowed El Al passengers to fly from Israel to Belgium on the Jewish Sabbath.

In early May, Sabena and City Bird were at odds over an arrangement struck by the Belgian upstart with a Congolese carrier, Lignes Aeriennes Congolaises (LAC). The deal allowed City Bird to operate the Brussels-Kinshasa route on behalf of the African airline. Sabena, which had that route covered, asked that City Bird pay BFR 350 million in damages. Sabena was also considering terminating its agreement with City Bird, claiming that the airline's actions violated the contract. As part of the agreement with City Bird, Sabena further noted that the airline was required to notify the Belgian carrier of any efforts that could impact its operations, either directly or indirectly. LAC, which took over operations from the defunct Air Zaire, agreed to a five-year contract with City Bird to fly passengers and cargo between Brussels and the former Belgian colony in competition with Sabena. The feud between City Bird and Sabena hit a new low in July when the low-cost carrier announced plans to file a complaint with the European Commission over unfair competition. The Belgian national airline countered it had a contract that required City Bird to notify Sabena of any competitive changes made. It had decided to file its complaint with the Belgian Civil Aviation Administration about the City Bird-LAC collaboration. A municipal arbitration court in Belgium was to resolve the matter.

Photo 117. Sabena MD-11 on final approach to Zurich on January 27, 2000. This MD-11 belonged to City Bird and was delivered to that airline on May 1, 1998. It is decked in Sabena colors as it was operated on behalf of the Belgian national airline. (Photo by Aero Icarus via Wikimedia)

Photo 118. Sabena MD-11 'OO-CTS' taxiing at Zurich on August 1, 1998. This aircraft was the second MD-11 delivered to City Bird. Delivery took place on April 10, 1998. It is seen here in Sabena colors, as it was also operated on behalf of the Belgian flag carrier. (Photo by Aero Icarus via Wikimedia)

The owner of City Bird and CEO, Victor Hassan, accused Sabena of having a monopolistic attitude contradictory to free-market competition. As an independent airline, he added that City Bird was entitled to do business and negotiate wet-lease contracts with other airlines. Sabena was in no mood to listen to City Bird's allegations because it was the low-cost carrier that had created the problem in the first place. In addition, Sabena sold 2.3% of its 11.2% stake in City Bird, a move that the national airline maintained was unrelated to the disagreement. City Bird countered that the sale, which caused City Bird's share price to collapse, was part of Sabena's strategy to punish City Bird. The case would be dragged through the legal system until the year 2000.

In April 2000, the decision by the court was made. The Brussels court ruled against City Bird in its quarrel with part-owner Sabena over the launch of services to the Democratic Republic of Congo in 1999. City Bird was made to pay Sabena nominal damages of just 1 Belgian franc. The court found no reason for the flag carrier to terminate its five-year contract with City Bird under which it was wet-leasing two City Bird Boeing MD-11s. Instead, the court decided that City Bird violated its agreement with Sabena by launching the Brussels-Kinshasa flights without notifying Sabena of its plans to operate a rival scheduled service. City Bird claimed its right to set its policies and emphasized that the Kinshasa route was operated on behalf of the Congo flag carrier LAC, for whom it wet-leased a Boeing 767-300ER. The irony was that City Bird had stopped its collaboration with LAC because of disagreements with the Congolese administration. Once more, City Bird claimed that Sabena sought vengeance by selling the majority of its 11.2% stake in City Bird, causing the value of the low-cost carrier's shares to fall. Sabena replied that the transaction was a legitimate financial transaction, not a spiteful one and that it still held a 2.99% stake in City Bird. City Bird's feathers were ruffled. Sabena was the clear winner.

After the fruitful code-sharing flights between Sabena and Delta Air Lines in 1998, the situation changed in 1999. In June, Air France and Delta joined forces to form a massive global alliance with a ten-year renewable term. Both passenger and cargo transportation was covered by the agreement between the French and American companies. Because Delta Air Lines had been Sabena and Swissair's partner for many years, the new alliance had a significant impact. The Swiss airline was dissatisfied with the situation and announced plans to sell its 4.6% investment in Delta Air Lines. Swissair had purchased the stock as a tactical investment in July 1989. Following the completion of the Delta-Air France alliance, Delta told Swissair and Sabena that the agreements made with those two airlines would be terminated in two years. The reason for such a time frame was that the contract stated that even if the arrangement were canceled, it would be in effect for at least one year. Delta CEO Leo F. Mullin reminded Swissair and Sabena that they had two years to join the Air France-Delta alliance. Sabena and Swissair decided that working with American Airlines would be better. That is what happened on November 16, 1999. The American Airlines deal was finally confirmed. Sabena and Swissair joined forces with American Airlines by signing a ten-year cooperation agreement. The agreement not only allowed for code-sharing flights, which corresponded to the pooling of reservations, and the use of different flight numbers for routes operated by the airlines in the alliance, but it also provided for a joint frequent flyer program. Sabena and the Swissair Group were glad to partner with American Airlines as their alliance with Delta would officially be over in 2000. American Airlines was a better option for Sabena and Swissair. It had a more extensive route network than Delta Air Lines, with many destinations throughout the USA and Latin America. Moreover, American Airlines flew to fewer destinations in Europe, which meant Sabena and Swissair would not need to compete with American Airlines in their home territory.

In 1999, Sabena's new Airbuses flew to the following destinations: Athens, Amsterdam, Dublin, Gothenburg, Hamburg, Helsinki, Lisbon, Lyon, Madrid, Manchester, Milan, Munich, Palma de Majorca, Venice, and Vienna. Sabena introduced Airbus A319 flights to Athens on September 1. Also, in September, Sabena started daily services from Brussels to Verona, with Avro RJ100s operated by its subsidiary Delta Air Transport (DAT). Then, on October 3, DAT inaugurated daily services for Sabena between Brussels and Seville, Spain. Regarding Germany, Sabena increased its daily flights from Brussels to Hanover from three to five and Hamburg from 4 to 6 in November. Also, a new, three times daily service between Brussels and Bremen began in September, operated by Schreiner Airways, a Dutch charter airline, under a franchise agreement with Sabena. In November, on the transatlantic route network, Sabena began daily services to Washington Dulles with its Airbus A330-300s. Delta Air Lines was due to codeshare the flights with Sabena, but with the American airline opting out of the agreement with Sabena and Swissair, the plan did not materialize. The good news was that, in that month, the US Department of Transport approved code-sharing flights between Sabena, and its new US partner airline, American Airlines, on services between Brussels and Washington DC, Boston and Chicago. Later in November, Sabena began flights from Brussels to Madras, India, using its Airbus A330-200s.

Sabena was ready to receive its new Airbuses in 1999. On March 2, Sabena's first Airbus A321 arrived in Brussels, shifting Sabena from an all-American fleet airline to an all-European one. The first of 33 Airbus Industrie A320s arrived in Belgium in November. As a result, Sabena operated all modern Airbus Industrie fly-by-wire aircraft, including single-aisle and long-haul widebodies. In 2002, Sabena planned to replace all its Boeing 737-200, 300, 400, and 500 short- to medium-haul airliners with 47 Airbus types. They included variants of the A319, A320, and A321 aircraft. Airbus A330s and A340s were already in use by the Belgian airline. It was the largest re-equipment initiative in Sabena's 75-year history. In November, Sabena announced it had phased out its last Boeing 747-329C 'OO-SGC' after the aircraft made its final flight on October 28, 1999. It left the Belgian carrier after that point. It would be the last US-built wide-body airliner in service with Sabena.

Photo 119. Sabena Airbus A319-112, 'OO-SSD' taxiing at Brussels Airport. Delivered to Sabena on October 22, 1999. Remarkably, it was put into operation the next day. (Photo by ETH Zurich)

Photo 120. Sabena Airbus A319-112, 'OO-SSK' on final approach at Zurich Kloten Airport on October 26, 2001. Delivered to the Belgian carrier on October 20, 2000. (Photo by Aero Icarus via Wikimedia)

Photo 121. Sabena A319-214 at a wet Manchester Airport on February 21, 2001. Delivered on January 12, 2001. (Photo by Ken Fielding via Wikimedia)

Photo 122. Sabena Airbus A320-214, 'OO-SNJ' at Paris Charles de Gaulle Airport on November 8, 2001. Delivered to Sabena on May 11, 2001. (Photo by Aero Icarus via Wikimedia)

Photo 123. Sabena Airbus A321-211, 'OO-SUC' in August 2001. This A321 still has the 'Qualiflyer Group' logo at the front of the aircraft, under the Sabena inscription. Delivered on May 4, 1999. It was Sabena's final A321.
(Photo by Aero Icarus via Flickr-Creative Commons 2.0)

Photo 124. Sabena A321-211 'OO-SUB' at Manchester Airport on February 9, 2000. Delivered on April 1, 1999.
(Photo by Ken Fielding via Wikimedia)

Nearing The End

2000-2001

2000

In February, Sabena announced it had made a loss of BFR 563 million in 1999 after registering its first profit in 40 years in 1998. The Belgian carrier attributed the outcome to high fuel prices and transatlantic overcapacity, and the cost of beginning its Airline Management Partnership with Swissair. Compared to 1998, operating profit was down 84% to BFR 597 million, with a gain of BFR 237 million after-tax. However, restructuring costs offset this, resulting in a net loss for the airline following a net profit of BFR 902 million in 1998. In April, the Swissair Group increased its stake in the Belgian airline Sabena to 85% after a public signing of the agreement by the Belgian government and the SAirGroup. It was the first time the Swissair Group had taken over the majority of a foreign airline. In addition, the Belgian state held an option to increase the stake in the Swissair Group to 5.5% if the country should sell the remaining 15% of its to Sabena shares later. Furthermore, the Belgian Government acquired a 3.3% stake in the Swissair Group with the signing of the agreement. The Belgian state also received the right to appoint a director to the Swissair Group's board and nominate Sabena's chairman and two directors. Moreover, the Swissair Group guaranteed Sabena's home carrier status in Belgium, and the Belgian airline retained its brand.

There was also the matter of Sabena's finances to consider. In October, the Belgian flag carrier was adamant about tightening its belt to improve its financial status. To begin with, it needed to save the equivalent of BFR 14.8 billion over the next year. It also had to lay off 500 full-time employees out of 12,000. Sabena's new CEO, Christoph Mueller, who took over for Swiss Paul Reutlinger, announced the measures to put the company back on track to profitability, especially since Sabena had lost more than BFR 3.37 billion in the first half of 2000 alone. According to Sabena's management, Sabena relied heavily on natural departures rather than on the sacking of hundreds of staff. In addition, the company intended to boost employee productivity by reorganizing work and encouraging adaptability. Sabena's management anticipated that most of the savings would come from management actions such as reducing direct flights to New York and Johannesburg. As a result, the two aircraft on those two routes would be able to be put on the market. Sabena had already grounded two A-340s and placed them for sale in June. The future of the airline's fleet size was in the line of fire, which was interesting, considering that in March 2000, Sabena firmed up its memorandum of understanding, placed in December 1999, for four new Airbus A340-300s, plus options on a further 4 A330s And A340s, whose deliveries were due to take place in 2001 and 2002.

In Europe, during June, Sabena and LOT Polish Airlines increased flights between Brussels and Warsaw to 4 per day. Sabena also added more flights on its Virgin Express operated services between Brussels and Rome, to 7 a day, and added two extra flights between Brussels and Marseilles, making it 5 per day. On the North Atlantic network, Sabena began daily services from Brussels to Dallas, Fort Worth, on August 6, using its Airbus A340-300s. These were code-sharing flights with American Airlines.

The Final Year

2001

Sabena held a crisis meeting of its shareholders, the Swissair Group, and the Belgian government on February 8 to avert a financial crisis threatening the airline's future. The meeting was convened after Sabena's worst performance in its long history, with the Belgian flag carrier losing an estimated BFR 7 billion in 2000. Sabena said that rising fuel prices, low yields, an unfavorable US dollar exchange rate, and higher-than-expected expenditures for replacing its all-Boeing fleet with Airbus planes manufactured in Europe significantly impacted its financial position. There were also state laws that put the carrier in a tight spot. Article 103 of Belgium's business practices statute required any company that had spent more than half of its capital to correct the issue either by declaring bankruptcy or recapitalizing. Fortunately, Sabena had a recovery plan in place, which included a BFR 30 billion recapitalization of the airline. One-third of the funds would come from the Swissair Group's BFR 6 billion investment and BFR 4 billion from the Belgian government. BFR 14 billion, or about half, had to come from the 'Blue Sky' savings plan. This plan notably called for the business to be recapitalized by BFR 10 billion (30 million euros) per year. The employee unions agreed to a systematic savings plan of BFR 2.1 billion (52 million euros). Sabena's unions opposed the latter despite Sabena's assurances that no jobs would be lost. This point was the primary focus of negotiations with the unions. After lengthy talks, the Belgian government and the Swissair Group agreed to provide Sabena with a financial injection of BFR 10 billion (250 million euros). In addition, the two significant shareholders announced a possible reassessment of their partnership simultaneously. After the unions accepted Sabena's restructuring plan, nothing now stood in the way. Once again, the Belgian airline was saved by the bell. However, it could not rely on intervention by its partners forever.

In May, with the financial storm that was crippling Sabena continuing unabated, the Belgian airline decided to delay the delivery of the last 15 Airbus A320s that it had ordered. With its future uncertain, it could not accept any additional planes. A Sabena representative insisted that the order was not canceled, but who was he to say? The bell was tolling for Sabena. The carrier had ordered 34 Airbus A 320s in total, of which 19 were operational and 15 were yet to be delivered. The Swissair Group was blamed for Sabena's difficult predicament with the aircraft purchases. The Swiss airline first advocated the acquisition of European Airbus planes to standardize their fleets and, in particular, to allow for mutual pilot training. On the other hand, the Belgian pilots fiercely protested the purchase of the aircraft, viewing it as an expensive move that went against Sabena's interests. With the delivery slowdown, all Sabena wanted was to get back into the black. It sounded like naive optimism. Mario Corti, the ex-CEO of Nestle's USA operations and the head of the Swissair Group, said that a decision on the Sabena 'situation' would come during the summer.

In June, he did what he preached. Mario Corti stated that the Swissair Group was not prepared to provide any further funds to Sabena. Swissair's CEO increased the pressure on Sabena's co-shareholder, the Belgium Government, by telling the Swiss press that he could not agree with Swissair's shareholders continuing to pay for the losses of other airlines. Corti was unimpressed with the Belgian Government. The Belgian Prime Minister, Guy Verhofstadt, said that Swissair had to fulfill its responsibilities under the Sabena contract. If that was not the case, he stated that the State reserved the right to take appropriate action. The Belgian government stepped in as Sabena's primary defender in its hour of need, ensuring that Swissair met its obligations to Sabena and its 11,000 employees. Mr. Verhofstadt demanded that Swissair honor its deal to purchase an 85% stake in Sabena, which was reconfirmed at the start of 2001. If Swissair did not comply, it would have to pay financial compensation.

In July, the Belgian government announced plans to file a slew of lawsuits against Swissair if negotiations over Sabena's future fell through. The threat came after the Belgian government rejected Swissair's proposal to get out of a pact to increase its ownership in the Belgian flag carrier from 49.5% to 85% by paying for a portion of a recovery plan. The threat worked. On July 2, Mario Corti and Swissair came up with a new offer for the Belgian government to assist Sabena in continuing its operations. According to the Swissair plan, the two shareholders would contribute BFR 12.3 billion (305 million euros) to Sabena. Swissair and the Belgian Government would need to inject a total of BFR 11 billion (275 million euros) into Sabena, with each party paying half of the cost.

Consequently, Swissair would grant a loan of BFR 1.2 billion (30 million euros), which would add up to BFR 12.3 billion. In addition, Swissair requested that it be released from its obligations under the agreement to increase its stake in Sabena to 85% and any future financial obligations related to Sabena. The offer did not go down well in Brussels. Swissair and the Belgian Government's disagreement over Sabena continued. Belgian Minister Rik Daems stated that the Swissair Group would be sued for breach of contract for BFR 30 billion (760 million euro) in damages, which was met with disappointment by the ailing Swiss carrier.

There was a breakthrough on July 17. The two parties agreed to inject BFR 17 billion (430 million euros) into Sabena. The Swissair Group would cover 60% of the cost, while the Belgian government would cover 40%. As a result, Swissair Group's 49.5% stake in Sabena would be preserved. The previous contract, signed on January 25, had been canceled by the new agreement. Swissair's management stated that the Swissair Group and the Belgian government were freed from other financial contributions. The funds would be paid out in four installments over two years, with the first installment due in October. Swissair was finally able to resolve the legal aspect of the case without any damage to its reputation. If the dispute remained unresolved, it would have taken on an embarrassing appearance for the Swiss airline, especially after the Belgian Government and Sabena filed suits for breach of contract by the Swissair group, which would have tainted the Swiss carrier's stellar reputation in the eyes of the public.

Sabena was not out of the woods yet. In September, the airline's Board held a crisis meeting with CEO Christoph Mueller, warning that the restructuring plan must be negotiated as a matter of urgency between the airline and the unions; if not, the airline would cease to exist. The reason for the stark warning by the CEO was due to the continuing deadlock between management, unions, and pilots, which threatened the survival of Sabena, in which Swissair also still held a 49.5% stake. These internal tensions were part of a context of an overall air transport crisis after the 911 attacks against the United States by Middle Eastern terrorists. Since that tragic event, Sabena had been losing BFR 40 million a day (1 million euro) because of the cancellation of the transatlantic flights, putting a massive dent in its finances.

In September, the situation within the company was unstable after management decided to take its conflict with the pilots to court. After that, a strike was announced on September 6, and the pilots walked out the next day. Sabena's management then brought an action for interim measures threatening the strikers and their associates with severe penalties. The case was taken to court by Sabena, and the pilots appealed this decision. The unions supported the pilots and warned that the right to strike by any Sabena employee was under threat. The unions had previously announced that they had reached an agreement with Sabena on the restructuring plan. However, the Pilots' Association refused to sign the deal.

The blow for Sabena came on October 1, when the Swissair Group, which was also in a dire financial situation, announced that it would not pay the money promised for the recapitalization of Sabena. Two days later, on October 3, Sabena's management requested the opening of a concordat procedure, which is a reconstruction agreement between the debtor (Sabena) and its creditors. The Belgian government agreed and provided Sabena with a one-month bridge loan of BFR 5 billion (125 million euros). In addition to the bridge loan, the Belgian government began seeking private partners to find a viable long-term solution for Sabena. The government focused on forming a smaller, trimmer Sabena, which would be successful because it was situated in Europe's capital, Brussels, whose airport welcomed more than 20 million passengers a year.

Later that month, the European Commission authorized the Belgian State to grant Sabena a bridging loan of BFR 5 billion. The airline was placed under judicial composition, by which a company in a state of insolvency (Sabena) files a judicial reorganization request to recover its economic activities and be discharged from its debts according to a reorganization plan. This action temporarily sheltered Sabena from its creditors. Still, Sabena no longer had the necessary liquidity to continue its operations since the desertion of the Swissair group, which should have participated in the Belgian carrier's recapitalization but dropped Sabena like a hot potato.

The end was near now for Sabena. On October 30, Sabena announced the temporary suspension of most of its flights to Canada and the United States. Nonetheless, the planes were nowhere near full, as bookings had fallen since Sabena's placement under the concordat procedure. As a result, daily flights to Chicago, Dallas, and Montreal were suspended in early November. Then on November 7, 2001, the inevitable end arrived. After 78 years of its fascinating existence, the national airline of Belgium was gone. The nation of Belgium, a country that valiantly weathered two world wars, was left with Sabena's carcass. At 9 pm on the night of November 7, the bankruptcy of the Belgian airline was officially confirmed by the chairman of Sabena's board of directors, Fred Chaffart. In one fell swoop, Sabena was erased forever.

Below is a very brief history of Sabena's subsidiary, Sobelair. The airline rode the highs and lows together with its parent company. Unfortunately, Sabena's bankruptcy also pushed Sobelair to the edge of a cliff and its inevitable fall.

Sobelair

Sobelair was founded on July 30, 1946, by former Sabena pilots to offer tourist flights from the eastern Congo going via the Nile route to Belgium. The initial capital of BFR 3.5 million was increased to BFR 18 million in October 1947 by contributions from active companies in the Belgian Congo. In 1949 these companies sold their shares to the national airline, Sabena.

In 1947, Sobelair became the first airline from Belgium to offer flights to Ethiopia. Initially, the aircraft were the small, rugged Douglas DC-3s. Later, they included the larger DC-4s and DC-6s. For safety reasons, the flights took place during the day. Passengers were provided with overnight hotel accommodation at the stopovers.

In 1949, Sobelair announced it had reached 5,296 flight hours for the year, representing a distance of 822,695 miles (1,324,000 km), against 4,752 flight hours 738,188 miles (1,188,000 km) in 1948. It also transported 3,295 passengers in 1949 compared to 2,940 in the previous year. As a result, operating revenues amounted to BFR 54 million against BFR 39 million in 1948.

In 1957, Sobelair established a domestic route network in the Congo with a fleet of three Cessna 310s to complement the primary Sobelair services from Belgium. However, Sobelair stopped the domestic Congo services in June 1962 due to the country's independence, and the Belgian supplemental airline withdrew from Africa for good.

In 1968, Sobelair took delivery of its first turbo-powered airliner, an F-27-400 Combi, in April 1968. With the F-27, Sobelair operated the Brussels-Cologne/Bonn flights twice a day for Sabena and the Brussels-Stuttgart route once a day. In addition, the Friendship also operated a nightly Brussels-Schiphol freight service.

In 1971, the Sabena subsidiary took delivery of three ex Sabena twin-engine Caravelle airliners and put them in operation. As a result, Sobelair would now be able to fly to destinations on behalf of Belgian travel companies offering all-inclusive flights, mainly to Southern Europe and North Africa. Due to the close collaboration between Sobelair and Sabena, the Caravelles would be maintained in the workshops of the Belgian flag carrier.

In 1973, Sobelair leased two Boeing 707-329 from Sabena, enabling the supplemental carrier to offer more capacity and supplement the four Caravelle 6Ns and one Caravelle III it had in its fleet. Then, in 1977, Sobelair ordered its third jetliner type, a Boeing 737-200. Again, it was not a hand down from Sabena; the aircraft came directly from Boeing brand new. It was delivered to the charter airline the following year.

Over the next five years, Sobelair's workforce increased to 130, and the Caravelles were replaced by its three Boeing 737-229A twin jets. Finally, in 1987, Sobelair placed an order for five 737-300s delivered in August 1987. Meanwhile, it also made a handsome profit of BFR 41.8 million.

In 1988, it was the turn of the Boeing 707s in the fleet to be withdrawn. They were all replaced by the 737s. Suddenly, Sobelair had become an all-Boeing 737 fleet comprised of two B-737-229As, one B-737-329, and one B-737-46B. The next type of airliner to join the fleet was the 767-300ER. In 1994 the first leased 767 was delivered, and two more followed two years later. By 1994, just like its parent, Sabena, Sobelair was in a bad financial state. The company saw customer bookings drop 24.9% to 688,000, followed by another decline in 1995 of 11.4% to 609,400. In 1996, Sobelair leased two Boeing 767s, a 767-328ER and a 767-33AER. The introduction of the two leased 767s brought good omens for the carrier, as it enjoyed a better year than the previous two. It saw its passenger boardings jump 10.4% to 1,353,000. Moreover, Sobelair's 767s were making money for the airline. In the summer of 1997, Sobelair's 767-300ER was used on Swissair's leisure flights from Zurich to San Francisco and Las Vegas. In December 1997, Sabena agreed to deploy Sobelair 767-300s three times a week to Dakar, Banjul, Bangkok, and Phuket for the summer season. In November 1998, Sobelair operated a weekly 767-300ER service to the Caribbean island of Curacao from Brussels. Then it obtained two new 767-300ERs from Flightlease, which replaced Sobelair's two older 767-300s.

In the end, Sabena's collapse led to the events that would finish off Sobelair. After being purchased by a group of investors in June 2002, which resulted in regular flights on the Brussels-Johannesburg route, Sobelair was sold to SN Brussels Airlines in early 2003. It would continue to operate charter flights while flying for SN Brussels Airlines. Unfortunately, Sobelair could not improve its financial situation. Consequently, the famed Belgian subsidiary of Sabena filed for bankruptcy protection in early January 2004. TUI Belgium expressed an interest in taking over Sobelair's aircraft and employees to launch its announced low-cost subsidiary Sunrise. That proposition failed, as the Brussels Commercial Court declared Sobelair insolvent on January 19, 2004.

SOBELAIR'S AIRLINERS

Photo 125. Sobelair Caravelle VIN, 'OO-SRB,' was delivered to Sabena on February 1, 1961. It was leased to Sobelair on April 1, 1971. Broken up in Brussels in January 1980. It is seen here next to Air France Caravelle III 'F-BJTL.' (Photo by Eduard Marmet)

Photo 126. This Sobelair Caravelle VIN 'OO-SRC' was delivered to Sabena on February 16, 1961, and leased to Sobelair in April 1971. It was sold to Europe Aero Service in April 1978 and broken up in Perpignan, France, in May 1987. It is seen here in Munich in 1972. (Photo by Norbert Kröpfl)

Photo 127. Sobelair Caravelle VIN 'OO-SRB' is seen here in Munich in May 1976. It was with Sobelair until January 1980. (Photo by Norbert Kröpfl)

Photo 128. This is Sobelair 707-373C 'OO-SBU' at Brussels. Sobelair purchased it on January 5, 1979. In January 1990, it was sold to Florida West Airlines with registration N760FW. (Photo from 'www.douglasdc8.com')

Photo 129. Sobelair 707-329C 'OO-SJM' landing at Brussels. This aircraft was delivered to Sabena on June 17, 1969. Sobelair operated it for a while in the late 1970s. (Photo from 'www.douglasdc8.com')

Photo 130. This is another shot of Sobelair 'OO-SBU' at Athens Hellinikon Airport in July 1983. (Photo by Richard Vandervord)

Photo 131. Sobelair 707-329C 'OO-SJM' is seen here at London Heathrow Airport in March 1985. (Photo by Richard Vandervord)

Photo 132. Sobelair 'OO-SBT' in April 1987. Delivered to Sobelair on November 20, 1979. It was sold to Avant Airlines on October 27, 1997. (Photo by Udo K. Haafke)

Photo 133. Sobelair 737-229 Advanced 'OO-SBS' at Athens in June 1985. Delivered to Sobelair on August 14, 1979. It ended up with the Italian airline Air One in January 1999. Withdrawn from use at Pescara on November 30, 2010. (Photo by Richard Vandervord)

Photo 134. Sobelair 737-229 Advanced 'OO-SBT' coming in to land at Brussels Airport. (Photo from 'www.douglasdc8.com')

Photo 135. Sobelair 737-229 Advanced 'OO-SBT' at Athens in July 1983. Notice that this aircraft was operated by Air Belgium. (Photo by Richard Vandervord)

Photo 136. 737-229 Advanced 'OO-SBQ.' It was delivered to Sobelair on August 25, 1978. It was withdrawn from service on September 1, 1997, and sold to Avant Airlines on September 30, 1997. (Photo from 'www.douglasdc8.com')

Photo 137. 767-33AER, 'OO-SBY,' was the first wide-body airliner in Sobelair's service. It was leased by the airline on June 24, 1994. It was returned to the lessor on June 28, 2000. (Photo from 'www.douglasdc8.com')

Photo 138. This 767-3BGER, 'OO-SLS,' was delivered to Sobelair on November 22, 2000, and stayed with the supplemental carrier until its demise on January 19, 2004. (Photo by Richard Vandervord)

Photo 139. This is another photo of 'OO-SLS' on its final approach to London Heathrow in October 2003. (Photo by Richard Vandervord)

Creative Commons & Public Domain Photo Credits

Photo 1. https://commons.wikimedia.org/wiki/File:Portrait_of_Albert_I_of_Belgium_(cropped).jpg (original author: **Richard N. Speaight**, 1875-1938) Public Domain photo.

Photo 2. **ETH-Bibliothek Zürich**, Bildarchiv / Fotograf: Unbekannt / Ans_05338-02-124-AL-FL / Public Domain Mark

Photo 4. **Flightglobal** (https://commons.wikimedia.org/wiki/File:Westland_Wessex_plane.jpg), (Darkened and enhanced by Theodore Gianna), https://creativecommons.org/licenses/by-sa/4.0/legalcode

Photo 5. **Public Domain photo**. https://commons.wikimedia.org/wiki/File:R.a._-_Savoia_Marchetti_S-73P_C-N.jpg. Original author unknown. (lightened, smoothed out, sharpened, enhanced by Theodore Gianna)

Photo 6. **ETH-Bibliothek Zürich**, Bildarchiv/Stiftung Luftbild Schweiz / Fotograf: Swissair / LBS_SR01-02848 / CC BY-SA 4.0 (Brightened, sharpened by Theodore Gianna)

Photo 7. **ETH-Bibliothek Zürich**, Bildarchiv/Stiftung Luftbild Schweiz / Fotograf: Swissair / LBS_SR01-04433-01 / CC BY-SA 4.0 (colorized, exposure, denoise, sharpened by Theodore Gianna)

Photo 8. **ETH-Bibliothek Zürich**, Bildarchiv/Stiftung Luftbild Schweiz / Fotograf: Swissair / LBS_SR01-02858 / CC BY-SA 4.0 (colorized, lightened, denoise, sharpened by Theodore Gianna)

Photo 9. **ETH-Bibliothek Zürich**, Bildarchiv / Fotograf: Metzger, Jack / Com_L06-0267-0003-012 / CC BY-SA 4.0 (colorized, lightened, denoise, sharpened by Theodore Gianna) Sky photo: by **mrsiraphol** via freepik.com

Photo 10. **ETH-Bibliothek Zürich**, Bildarchiv / Fotograf: Metzger, Jack / Com_L06-0267-0018-0002 / CC BY-SA 4.0 (colorized, sharpened, smoothed, lightened by Theodore Gianna) Sky photo from **rawpixel.com**.

Photo-11..**Ken-Fielding**/https://www.flickr.com/photos/kenfielding (https://commons.wikimedia.org/wiki/File:OO-BIA_DH114_Heron_1_Sabena_RTM_10JAN68_(6812669405).jpg), „OO-BIA DH114 Heron 1 Sabena RTM 10JAN68 (6812669405)", Colorized, sharpened, enhanced by Theodore Gianna, https://creativecommons.org/licenses/by-sa/3.0/legalcode

Photo 13. **ETH-Bibliothek Zürich**, Bildarchiv/Stiftung Luftbild Schweiz / Fotograf: Swissair / LBS_SR01-05989 / CC BY-SA 4.0 (colorized, cropped, lightened by Theodore Gianna)

Photo.14..**Ken-Fielding**/https://www.flickr.com/photos/kenfielding (https://commons.wikimedia.org/wiki/File:OO-CTM_DC-6B_Sabena_LPL_25NOV64_(5574952641).jpg), „OO-CTM DC-6B Sabena LPL 25NOV64 (5574952641)", Sharpened, enhanced color, lightened by Theodore Gianna, https://creativecommons.org/licenses/by-sa/3.0/legalcode.

Photo 15. **ETH-Bibliothek Zürich**, Bildarchiv / Fotograf: Furrer, Bernhard / SIK_02-07-0125 / CC BY-SA 4.0 (colorized, lightened by Theodore Gianna) Sky photo by **Maddison McMurrin** via unsplash.com

Photo 16. Photo by **Harry Pot (Anefo), via Nationaal Archief, Netherlands**. CC0 license. Inventory File Number 905-7334.

Photo 17. **ETH-Bibliothek Zürich**, Bildarchiv/Stiftung Luftbild Schweiz / Fotograf: Swissair / LBS_SR03-07169 / CC BY-SA 4.0 (cropped, colorized, sharpened, lightened, sky photo added, by Theodore Gianna) Sky photo by: **Jenna Duffy** via unsplash.com

Photo 18. Photo by **Duinen (Anefo), via Nationaal Archief, Netherlands**. CC0 license. Inventory File Number 905-8741. Sky photo by **SHTTEFAN** via unsplash.com

Photo 19. **RuthAS** (https://commons.wikimedia.org/wiki/File:Douglas_DC-7C_OO-SFC_Sabena_RWY_18.08.62_edited-3.jpg), „Douglas DC-7C OO-SFC Sabena RWY 18.08.62 edited-3", Sharpened, brightened, enhanced color by Theodore Gianna, https://creativecommons.org/licenses/by/3.0/legalcode.

Photo 20. **Maurice Luyten** (https://commons.wikimedia.org/wiki/File:1958_SABENA_Fifties_Maurice_Luyten.jpg), cropped, brightened, enhanced color by Theodore Gianna, https://creativecommons.org/licenses/by/4.0/legalcode.

Photo 22. **ETH-Bibliothek Zürich**, Bildarchiv/Stiftung Luftbild Schweiz / Fotograf: Swissair / LBS_SR04-055843 / CC BY-SA 4.0 (cropped, enhanced color, brightened sharpened)

Photo.25..**Ken-Fielding**/https://www.flickr.com/photos/kenfielding (https://commons.wikimedia.org/wiki/File:OO-SCO_CV-440_Metropolitan_Sabena_OST_14SEP63_(6794839853).jpg), „OO-SCO CV-440 Metropolitan Sabena OST 14SEP63 (6794839853)", Sharpened, brightened, enhanced color by Theodore Gianna, https://creativecommons.org/licenses/by-sa/3.0/legalcode (enhanced color, brightened, sharpened)

Photo 28. **Robert-Huhardeaux**. Sabena Sikorsky S-58 at Brussels National Airport in 1962. https://www.flickr.com/photos/52086447@N00/17390107321/in/photolist-suGR3z-sbprJ8-rxJ7D5-2jQpQTA.
CC BY-SA 2.0 - https://creativecommons.org/licenses/by-sa/2.0/Photo

Photo 39. https://commons.wikimedia.org/wiki/File:Evacuatie_uit_Belgisch_Congo_te_Brussel_vliegveld_Zavente,_Bestanddeelnr_911-3490.jpg (original author: **Herbert Behrens** of Anefo, via Nationaal Archief of The Netherlands). Public Domain photo.

Photo 40. https://commons.wikimedia.org/wiki/File:Evacuatie_uit_Belgisch_Congo_te_Brussel_vliegveld_Zavente,_Bestanddeelnr_911-3493.jpg (original author: **Herbert Behrens** of Anefo, via Nationaal Archief of The Netherlands). Public Domain photo.

Photo 47. **ETH-Bibliothek Zürich**, Bildarchiv/Stiftung Luftbild Schweiz / Fotograf: Swissair Photo AG / LBS_L1-672785 / CC BY-SA 4.0 (cropped sharpened enhanced color by Theodore Gianna)

Photo 48. **ETH-Bibliothek Zürich**, Bildarchiv/Stiftung Luftbild Schweiz / Fotograf: Swissair / LBS_SR04-055433 / CC BY-SA 4.0 (cropped sharpened enhanced color by Theodore Gianna)

Photo-50.**Ken-Fielding**/https://www.flickr.com/photos/kenfielding (https://commons.wikimedia.org/wiki/File:OO-SRC_SE.210_Caravelle_6_Sabena_LPL_26NOV64_(6151271288).jpg), „OO-SRC SE.210 Caravelle 6 Sabena LPL 26NOV64 (6151271288)", Sharpened, enhanced color, lightened, https://creativecommons.org/licenses/by-sa/3.0/legalcode

Photo 66. **Clemens Vasters** from Viersen, Germany (https://commons.wikimedia.org/wiki/File:Royal_Military_Museum,_Brussels_-_The_Sabena_Concorde_that_never_was_(11449229885).jpg), „Royal Military Museum, Brussels - The Sabena Concorde that never was (11449229885)", https://creativecommons.org/licenses/by/2.0/legalcode

Photo 84. **Pedro Aragão** (https://commons.wikimedia.org/wiki/File:Embraer_EMB-121_Xingu,_Sabena_-_Belgian_Aviation_School_JP93097.jpg), „Embraer EMB-121 Xingu, Sabena - Belgian Aviation School JP93097", Sharpened, enhanced color, lightened by Theodore Gianna, https://creativecommons.org/licenses/by-sa/3.0/legalcode

Photo 86. **Michiel Hendryckx** (https://commons.wikimedia.org/wiki/File:Herman_de_croo_675.jpg), „Herman de croo 675", https://creativecommons.org/licenses/by-sa/3.0/legalcode.

Photo 111. **ETH-Bibliothek Zürich**, Bildarchiv/Stiftung Luftbild Schweiz / Fotograf: Swissair / LBS_SR04-013363 / CC BY-SA 4.0 (lightened, balanced colors by Theodore Gianna)

Photo 112. **Aero Icarus** from Zürich, Switzerland (https://commons.wikimedia.org/wiki/File:52aw_-_Sabena_Airbus_A330-301;_OO-SFO@ZRH;27.02.1999_(5134755347).jpg), „52aw - Sabena Airbus A330-301; OO-SFO@ZRH;27.02.1999 (5134755347)", (enhanced color, lightened), https://creativecommons.org/licenses/by-sa/2.0/legalcode

Photo 113. **ETH-Bibliothek Zürich**, Bildarchiv/Stiftung Luftbild Schweiz / Fotograf: Swissair / LBS_SR04-013577 / CC BY-SA 4.0 (brightened, sharpened by Theodore Gianna)

Photo 116. **Bob Adams** from George, South Africa (https://commons.wikimedia.org/wiki/File:City_Bird_B767-33A-ER_OO-CTA_(6781635777).jpg), „City Bird B767-33A-ER OO-CTA (6781635777)", Sharpened, reduced noise, enhanced color, lightened by Theodore Gianna, https://creativecommons.org/licenses/by-sa/2.0/legalcode

Photo 117. **Aero Icarus** from Zürich, Switzerland (https://commons.wikimedia.org/wiki/File:81ay_-_Sabena_MD-11;_OO-CTC@ZRH;27.01.2000_(5134756427).jpg), „81ay - Sabena MD-11; OO CTC@ZRH;27.01.2000 (5134756427)", Sharpened, enhanced color, lightened by Theodore Gianna, https://creativecommons.org/licenses/by-sa/2.0/legalcode

Photo 118. **Aero Icarus** from Zürich, Switzerland (https://commons.wikimedia.org/wiki/File:34bb_-_Sabena_MD-11;_OO-CTS@ZRH;07.08.1998_(5601117539).jpg), „34bb - Sabena MD-11; OO-CTS@ZRH;07.08.1998 (5601117539)", Sharpened, enhanced color, lightened by Theodore Gianna, https://creativecommons.org/licenses/by-sa/2.0/legalcode

Photo 119. **ETH-Bibliothek Zürich**, Bildarchiv/Stiftung Luftbild Schweiz / Fotograf: Swissair / LBS_SR04-013578 / CC BY-SA 4.0

Photo 120. **Aero Icarus** from Zürich, Switzerland (https://commons.wikimedia.org/wiki/File:157az_-_Sabena_Airbus_A319-112,_OO-SSK@ZRH,26.10.2001_-_Flickr_-_Aero_Icarus.jpg), „157az - Sabena Airbus A319-112, OO-SSK@ZRH,26.10.2001 - Flickr - Aero Icarus“, Sharpened, enhanced color, brightened by Theodore Gianna, https://creativecommons.org/licenses/by-sa/2.0/legalcode

Photo 121. **Ken Fielding**/https://www.flickr.com/photos/kenfielding (https://commons.wikimedia.org/wiki/File:OO-SNG_A320-212_Sabena_MAN_21FEB01_(6834430970).jpg), „OO-SNG A320-212 Sabena MAN 21FEB01 (6834430970)“, Sharpened, enhanced color, lightened, smoothed out, https://creativecommons.org/licenses/by-sa/3.0/legalcode

Photo 122. **Aero Icarus** from Zürich, Switzerland (https://commons.wikimedia.org/wiki/File:145gn_-_Sabena_Airbus_A320-214,_OO-SNJ@CDG,11.08.2001_-_Flickr_-_Aero_Icarus.jpg), „145gn - Sabena Airbus A320-214, OO-SNJ@CDG,11.08.2001 - Flickr - Aero Icarus“, Sharpened, enhanced color, lightened by Theodore Gianna, https://creativecommons.org/licenses/by-sa/2.0/legalcode

Photo 123. **Aero Icarus** from Zürich, Switzerland. (https://www.flickr.com/photos/aero_icarus/5702857332/in/photolist-hyigHW-8NrLqn-9sHDpd-248RZsk-aps1jd-9FWCNq-drT5p1-9x4q5r-8X3TcL-dup3Uq-fK3u2Z-8PN2Xj-M9BiFo-8P3Rnn-8PJY8n-8WZU8B-9vPPLf-8trstf-8PJZZv-8PN35L-8PN54S-Mqv4Jy-8PJXNv-XBZJ6m-T8mrAN-RdzdkP-UEDJ5P-dkZ9T1-8YQiLX-aps1MW-8P3Qvc-8PJXSB-PRT5y1-ms2meH-nMr8XR-8PJXCB-8Smocb-GZMB4M-qdMHWo-UB2NuS-QeKPPQ-9x7riJ-8P3NnM-8YQ6e8-8P3MWr-9x7psQ-8PJXFg-8PJXJn-NZ98nX-9x4o6M) Sharpened, enhanced color, lightened by Theodore Gianna https://creativecommons.org/licenses/by-sa/2.0/legalcode

Photo 124. **Ken Fielding**/https://www.flickr.com/photos/kenfielding (https://commons.wikimedia.org/wiki/File:OO-SUB_A321-211_Sabena_MAN_09FEB00_(5898679420).jpg), „OO-SUB A321-211 Sabena MAN 09FEB00 (5898679420)“, Sharpened, enhanced color, lightened, smoothed out by Theodore Gianna, https://creativecommons.org/licenses/by-sa/3.0/legalcode

Back cover photo: ETH-Bibliothek Zürich, Bildarchiv/Stiftung Luftbild Schweiz / Fotograf: Swissair / LBS_SR01-02958 / CC BY-SA 4.0

INDEX

Made in the USA
Las Vegas, NV
03 November 2026

57421893R00124